# BestMasters

Mit „BestMasters" zeichnet Springer die besten Masterarbeiten aus, die an renommierten Hochschulen in Deutschland, Österreich und der Schweiz entstanden sind. Die mit Höchstnote ausgezeichneten Arbeiten wurden durch Gutachter zur Veröffentlichung empfohlen und behandeln aktuelle Themen aus unterschiedlichen Fachgebieten der Naturwissenschaften, Psychologie, Technik und Wirtschaftswissenschaften. Die Reihe wendet sich an Praktiker und Wissenschaftler gleichermaßen und soll insbesondere auch Nachwuchswissenschaftlern Orientierung geben.

Springer awards "BestMasters" to the best master's theses which have been completed at renowned Universities in Germany, Austria, and Switzerland. The studies received highest marks and were recommended for publication by supervisors. They address current issues from various fields of research in natural sciences, psychology, technology, and economics. The series addresses practitioners as well as scientists and, in particular, offers guidance for early stage researchers.

More information about this series at https://link.springer.com/bookseries/13198

Jonathan Steinberg

# Extensions and Restrictions of Generalized Probabilistic Theories

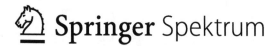 **Springer** Spektrum

Jonathan Steinberg
Lennestadt, Germany

ISSN 2625-3577          ISSN 2625-3615   (electronic)
BestMasters
ISBN 978-3-658-37580-5          ISBN 978-3-658-37581-2   (eBook)
https://doi.org/10.1007/978-3-658-37581-2

Responsible Editor: Marija Kojic
This Springer Spektrum imprint is published by the registered company Springer Fachmedien
Wiesbaden GmbH part of Springer Nature.
The registered company address is: Abraham-Lincoln-Str. 46, 65189 Wiesbaden, Germany

# Abstract

Generalized probabilistic theories (GPTs) allow us to write quantum theory in a purely operational language and enable us to formulate other, vastly different theories. As it turns out, there is no canonical way to integrate the notion of subsystems within the framework of convex operational theories. Sections can be seen as generalization of subsystems and describe situations where not all possible observables can be implemented. We discuss the mathematical foundations of GPTs using the language of Archimedean order unit spaces and investigate the algebraic nature of sections. This includes an analysis of the category theoretic structure and the transformation properties of the state space. Since the Hilbert space formulation of quantum mechanics uses tensor products to describe subsystems, we show how one can interpret the tensor product as a special type of a section. In addition we apply this concept to quantum theory and compare it with the formulation in the algebraic approach. Afterwards we give a complete characterization of low dimensional sections of arbitrary quantum systems using the theory of matrix pencils. In addition, we combine the notion of sections with the dynamics in a GPT and consider the implications for quantum theory. As an application we introduce Spekkens' toy model, a hidden variable theory which mimics many of the features of quantum theory. We show that this model cannot be obtained as a section of a qubit system, but emerges naturally as a section of the Kadison representation what can be interpreted as an eight-level quantum system.

# Contents

# Introduction

1

Quantum theory appears in many scientific fields and is one of the most success-ful physical theories. As one would expect, there is a wide range of experimental verification with high precision and without loopholes [1]. Thus it seems, that the theoretical and mathematical framework of quantum theory is an assured knowl-edge and well understood. Though the meaning and the origin of the counterintuitive Hilbert space formalism are still widely considered to be difficult to grasp. Thus, the goal is to find a compelling set of axioms that single out quantum theory from other possible theories. Finding such a set of axioms is a problem starting with the work of Birkhoff and von Neumann [2] and continuing through the works of Mackey [3], Ludwig [4] and Piron [5] and the tradition of quantum logic [6]. But to understand what distinguishes quantum mechanics from alternative theories, one needs a framework that allows us to describe a broad range of physical theories, including quantum and classical theories, but also contain more exotic ones. Con-structing this framework is a highly non-trivial task, where one is compelled to make fundamental choices as what is considered to be a general property of the theories and what specific. In particular, the specific part can serve as an candidate for an axiom that identifies quantum theory. With the raise of quantum information and its emphasis on simple operational setups, frameworks that focus on a strictly operational formulation of physical theories gained popularity. Following [7], we call those frameworks that can describe features of experimental probabilities while remaining noncommittal about the realization of the chosen structures generalized probabilistic theories (GPTs). In particular, the very flexible mathematical frame-works of GPTs turned out to be very fruitful for fundamental investigations of information-theoretical aspects of quantum theory [8–10]. Two representatives of GPTs are convex operational theories (COT) [11] and operational-probabilistic the-ories (OPT) [12, 13]. COP describe preparations, transformations and measurements by elements of suitable convex sets in an ambient vector space and allow by virtue

J. Steinberg, *Extensions and Restrictions of Generalized Probabilistic Theories*, BestMasters, https://doi.org/10.1007/978-3-658-37581-2_1

of dual forms the calculation of probabilities. OPTs can be seen a descendant of the category-theoretic framework [14, 15]. The main idea of this approach is that the composition of systems and processes is fundamental to the operational structure of a theory and that one can talk about information processing without having the notion of probabilities. However probabilistic predictions of an operational theory are often subject of interest. Thus OPTs join the compositional and the probabilistic features. In difference to the COT framework, OPTs also allow the description of theories, where the state space is not convex.

Below, we will mainly focus on the convex approach. Starting from the set of all convex operational theories, it is also possible to uniquely determine quantum mechanics by specifying some of its characteristic properties [16, 17]. One requirement for a fundamental physical theory is the possibility of the description of subsystems [18–20]. In difference to operational-probabilistic theories where the notion of composing is an inherent part of the framework, the notion of subsystems and composition in convex operational theories turn out to be rather complicated. To be more precise, the convex structure do not allow the decomposition of a system in subsystems in terms of a tensor product. One attempt to resolve this problem is to introduce the concept of sections [21]. A section can be seen as a restriction of the accessible physical quantities of a system, i.e., not the whole set of observables can be implemented. However, the notion of sections was not motivated by a theoretical description of subsystems, but rather by a more abstract question. If one assumes that quantum mechanics is not a fundamental description of nature, why is quantum mechanics such a good effective theory? As it turns out [21], if the physically implementable measurements are chosen from a typical subset, then the bipartite correlations arising in GPTs can be modeled, to a high degree of precision, by those of quantum mechanics.

The aim of this thesis is to give a systematic introduction to the abstract mathematical description of GPTs and to present an interpretation of sections as subsystems in this framework. For this purpose, Chapter 2 deals with the mathematical notions needed in the subsequent parts. This includes an introduction to Archimedean ordered vector spaces, categories, completely positive maps and convex polytopes. In Chapter 3, we summarize the description of physical systems in terms of GPTs and present some properties that that can be used for a characterization of GPTs, e.g., cloning, broadcasting and teleportation protocols. Further, we formulate classical probability theory, quantum mechanics and the PR-box in terms of this framework. Chapter 4 contains the definition of section and carves out its main physical and mathematical properties. In particular, it gives a characterization of how the state space transforms if the cone of observables undergoes a section. Further, we compare the new concept of sections with the notion of subsystems in the algebraic approach to quantum theory [22–25]. As it turns out, understanding the geometry of sections or the equivalent geometry of the transformed state space is a hard task. Thus, Chapter 5 deals with the geometrical characterization of subspaces of a state space arising from a $\ell$-level quantum system state space. Namely, we give necessary conditions for a section of the state space to be a polytope. In conclusion, we analyze how the geometry of a section behaves, if the whole system undergoes a evolution or a process.

# Mathematical Preliminaries

## 2.1 Basic Concepts of Vector Spaces

In this section we introduce the concepts of cones, ordered vector spaces and states. In the notation we are mainly following [26]. Note that the concepts introduced in the following paragraphs do not only apply to finite dimension vector spaces, but reveal its full power in the context of infinite dimension spaces, e.g., function spaces. During this section we write $\mathbb{F} = \{\mathbb{R}, \mathbb{C}\}$ to indicate that the propositions hold for the real as well as for the complex case.

### 2.1.1 Minkowski Functionals

**Definition 2.1** Let $E$ be a $\mathbb{F}$-vector space. We say that a subset $A \subset E$ is an absorbing set for $E$, if

$$E = \bigcup_{t>0} tA \tag{2.1}$$

For an absorbing set $A \subset E$ we define the Minkowski functional or gauge functional $p_A : E \to [0, \infty)$ via

$$p_A(x) := \inf\{t > 0 \,|\, x \in tA\} \quad , \quad x \in E \tag{2.2}$$

Further $A \subset E$ is called absolutely convex if

$$\forall x, y \in A \; \forall s, t \in \mathbb{F} : |s| + |t| \leq 1 \Rightarrow sy + ty \in A \tag{2.3}$$

© The Author(s), under exclusive license to Springer Fachmedien Wiesbaden GmbH, part of Springer Nature 2022
J. Steinberg, *Extensions and Restrictions of Generalized Probabilistic Theories*, BestMasters, https://doi.org/10.1007/978-3-658-37581-2_2

**Theorem 2.2** Let $E$ be a $\mathbb{F}$-vector space and $C \subset E$ a convex, absorbing set. Then the Minkowski functional $p := p_C : E \to [0, \infty)$ is sublinear. Is $C$ absolutely convex, then $p_C$ is a seminorm.

*Proof.* Clearly $0 \leq p_C(x) < \infty$. For $x \in E$ one can find $t, x > 0$ such that $tx \in C$ and $s(-x) \in C$. Since $C$ is also convex, we have $0 \in C$ and $p_C(0) = 0$. For $\lambda > p(x)$ and $\mu > p(y)$ we have $x \in \lambda C$ and $y \in \mu C$ due to the convexity of $C$ and $0 \in C$. From this we obtain

$$\frac{1}{\lambda + \mu}(x + y) = \frac{\lambda}{\lambda + \mu}\frac{x}{\lambda} + \frac{\mu}{\lambda + \mu}\frac{y}{\mu} \in C \tag{2.4}$$

what implies $x + y \in (\lambda + \mu)C$. Therefore we have $p(x + y) \leq \lambda + \mu$ what implies the triangle inequality $p(x + y) \leq p(x) + p(y)$. For homogeneity let $t > 0$ and so $tx \in t\lambda C$ what implies $p(tx) \leq tp(x)$. For $t > 0$ also the opposite is true $p(x) \leq \frac{1}{t}p(tx)$ and we obtain $p(tx) = tp(x)$ for $t \geq 0$. Is in addition $C$ absolutely convex and $\lambda > p(x)$ we have for $\alpha \in \mathbb{F}$ that $\alpha x \in \alpha \lambda C = |\alpha|\lambda C$. Thus $p(\alpha x) \leq |\alpha|p(x)$. With the same argument as above we have $p(\alpha x) = |\alpha|p(x)$. □

## 2.1.2   The Hahn-Banach Theorem

In this section we present a version of the Hahn-Banach theorem. As an important application a separation theorem for topological vector spaces is derived. Since this thesis mostly deals with real vector spaces, we only state the Hahn-Banach theorem for the real case.

**Definition 2.3** Let $V$ be a $\mathbb{R}$-vector space. A map $p : V \to \mathbb{R}$ is called a sublinear functional if

1)  $p(x + y) \leq p(x) + p(y)$ , $x, y \in V$
2)  $p(tx) = tp(x)$ , $t \in \mathbb{R}^+$ , $x \in V$

Clearly seminorms and norms are examples for sublinear functionals.

**Theorem 2.4** ([27]). Let $W$ be a $\mathbb{R}$-vector space and $V \subset W$ a linear subspace. Further let $p : W \to \mathbb{R}$ be a sublinear functional and $f_0 : V \to \mathbb{R}$ linear with $f_0(x) \leq p(x)$ for all $x \in V$. Then there exists a functional $f : W \to \mathbb{R}$ such that

$$f|_V = f_0 \quad \text{and} \quad -p(-x) \leq f(x) \leq p(x) \ , \ x \in W \tag{2.5}$$

Let $X$ be a normed space $x \in X$ and $x' \in X'$ a functional. In analogy to inner products we write

$$\langle \cdot, \cdot \rangle : X \times X' \to \mathbb{F} \ , \quad \langle x, x' \rangle := x'(x) \tag{2.6}$$

Thanks to Theorem 2.4 we can find a functional $x' \in X'$ such that $||x'|| = 1$ and $|\langle x, x' \rangle| = ||x||$. Clearly we have $||x|| = \max\{|\langle x, x' \rangle| : x' \in X' , ||x'|| = 1\}$ for all $x \in X$. Thus we can identify elements of a Banach space with continuous functionals on its dual space and by reason of (2.6) both norms are the same. Due to technical reasons this thesis mostly deals with spaces where we obtain all continuous functionals on the dual by this construction. To do so, we define the canonical embedding $\iota = \iota_X : X \to X''$ of a norm space into its double dual $X'' := \left( X' \right)'$ via

$$\langle x', \iota x \rangle := \langle x, x' \rangle \quad \text{where} \quad x \in X \ , \ x' \in X' \tag{2.7}$$

It follows directly from (2.6) that is an isometry. Further we call a normed space $X$ reflexive if the canonical embedding into its double dual is surjective. If $\dim(X) < \infty$ then $\dim\left( X'' \right) < \infty$ and therefore $\iota_X$ is an isometric isomorphism. This allows us to identify $X \cong X''$. Except from this finite dimensional case it is important to note that an arbitrary normed space $X$ is not necessarily isomorphic or isometric to $X''$. One can construct a Banach space [28] $J$ with $J \cong J''$ but $\dim\left( J'' / \iota(J) \right) = 1$, thus not reflexive.

**Theorem 2.5** Let $X$ be a normed space over $\mathbb{R}$ or $\mathbb{C}$ and $\emptyset \neq D, B \subset X$ disjoint, convex sets.

1) Is $D$ open, then there exists $x' \in X'$ and $\gamma \in \mathbb{R}$ such that

$$\text{Re} \langle d, x' \rangle < \gamma \leq \text{Re} \langle b, x' \rangle \ , \ d \in D, b \in B \tag{2.8}$$

2) Is $D$ compact and $B$ closed, then there exists $x' \in X'$ and $\gamma_1 < \gamma_2 \in \mathbb{R}$ such that

$$\text{Re} \langle d, x' \rangle \leq \gamma_1 < \gamma_2 \leq \text{Re} \langle b, x' \rangle \ , \ d \in D, b \in B \tag{2.9}$$

*Proof.* Because any $z = x + iy \in \mathbb{C}$ can be written as $z = \operatorname{Re}(z) - i\operatorname{Re}(iz)$ we can restrict to the case $\mathbb{F} = \mathbb{R}$. We choose $d_0 \in D$. $b_0 \in B$ and define $x_0 = b_0 - d_0$. Hence the set $C := D - B + x_0$ is convex and open and since $0 \in C$ there exists a $\delta > 0$ such that $U_\delta(0) \subset C$. It follows that $C$ is an absorbing set for $X$ and by Theorem 2.2 the Minkowski functional $p = p_C$ is sublinear. By assumption we have $D \cap B = \emptyset$ and $x_0 \notin C$ and we have $p(x_0) \geq 1$. Consider the linear subspace $V_0 = \operatorname{span}(x_0)$ and define the linear functional $_0 : V_0 \to \mathbb{R}$ defined by $f_0(\alpha x_0) := \alpha$. Clearly

$$f_0(\alpha x_0) = \alpha \leq \alpha p(x_0) = p(\alpha x_0) \text{ for } \alpha \geq 0$$

$$f_0(\alpha x_0) = \alpha \leq 0 \leq p(\alpha x_0) \text{ for } \alpha \leq 0$$

From this we can conclude that $f_0 \leq p$ on $v_0$. By Theorem 2.4 one can find a functional $f : X \to \mathbb{R}$ with $f|_{V_0} = f_0$ and $-p(-x) \leq f(x) \leq p(x)$ for all $x \in X$. From $||x|| < \delta$ we can conclude that $x \in C \cap (-C)$, thus $|f(x)| \leq 1$ what implies $x' := f \in X'$ continuous. For $d \in D$ and $b \in B$ one has $\langle d, x' \rangle - \langle b, x' \rangle + 1 = \langle d - b + x_0, x' \rangle \leq p(d - b + x_0) < 1$ since $C$ is an open subset. Therefore $\langle d, x' \rangle < \langle b, x' \rangle$ and with $\gamma := \sup_{d \in D}\langle d, x' \rangle$ the claim 2.8 follows since $x'(D)$ is open. For the second claim let $D$ be compact and $B$ closed. If $d \in D$, then there exists $\epsilon := \epsilon_d > 0$ with $U_\epsilon(d) \cap B = \emptyset$. Due to the compactness of $D$ we can find $d_1, ..., d_r \in D$ such that

$$D \subset \bigcup_{j=1}^{r} U_{\epsilon_{d_j}}(d_j)$$

Define $\epsilon := \min\{\epsilon_{d_1}, ..., \epsilon_{d_r}\}$ and set $D_\epsilon := D + U_\epsilon(0)$. Now $D_\epsilon$ is open and convex with $D_\epsilon \cap B = \emptyset$. Because the first claim is already proven, we obtain $x' \in X'$ and $\gamma_2 \in \mathbb{R}$ such that $\langle x, x' \rangle < \gamma \leq \langle b, x' \rangle$ for $x \in D_\epsilon$ and $b \in B$. With $\gamma_1 := \max_{d \in D}\langle d, x' \rangle < \gamma_2$ the claim 2.9 follows.                    $\square$

There are many ways how one can define the concept of a topology on a set $x$, e.g., via neighbourhoods, via open sets, via closed sets or via the Kuratowski closure axioms. Because we are only interested in the application to vector spaces, we restrict to the definiton via open sets.

**Definition 2.6** Let $X$ be a set. By a topological space we mean an ordered pair $(X, \tau)$, where $\tau$ is a collection of subsets of $X$, satisfying the following axioms

1) The empty set and the set $X$ itself belongs to $\tau$
2) Any arbitrary union of members of $\tau$ belongs to $\tau$
3) The intersection of any finite number of members of $\tau$ belongs to $\tau$

A subset $S \subset X$ is said to be closed if $X \setminus S$ is open. The closure of $S$, denoted by $\overline{S}$ is the intersection of all closed subspaces of $X$ containing $S$. A subset $M \subset X$ of a topological space $(X, \tau)$ is called a neighbourhood for a point $x \in M$ if there exists $U \in \tau$ such that $x \in U \subset M$. We call $x, y \in X$ separated by neighbourhoods if there exists a neighbourhood $U$ of $x$ and a neighbourhood $V$ of $y$ such that $U \cap V = \emptyset$. The space $X$ is called a Hausdorff space or $T_2$ space if all distinct points in $X$ are pairwise neighbourhood-separable. Probably the simplest way to obtain new topological space is from taking subsets of other spaces. If $X$ is a topological space and $S \subset X$ is an arbitrary subset, we define the subspace topology on $S$ by declaring a subset $U \subset S$ to be open in $S$ if and only if there exists an open subset $V \subset X$ such that $U = V \cap S$. Any subset of $X$ endowed with the subspace topology is said to be a subset of $X$. A map $f : X \to Y$ with $X$ and $Y$ are topological spaces, is said to be continuous if for every open subset $U \subset Y$, the preimage $f^{-1}(U)$ is open in $X$. Further $f : X \to Y$ is called a topological embedding if it is a homeomorphism onto its image $f(X) \subset Y$ in the subspace topology.

At some point it turns out, that the collection of subsets $\tau$ is too big. This leads to the concepts of a base and a subbase. In particular bases are useful since many properties of a topology can be reduces to statements about its basis and because many topologies are most easily defined in terms of a base which generates them.

**Definition 2.7** Let $(X, \tau)$ be a topological space. A basis is a collection $\mathcal{B}$ of subsets of $X$ with the properties

1) The base elements cover $X$ i.e., if $\mathcal{C} : \{U_i : i \in I\}$ with $I$ index set, we say that $\mathcal{C}$ covers $X$ if $X \subset \cup_{i \in I} U_i$
2) For $\mathcal{B}_1, \mathcal{B}_2$ base elements define $V := \mathcal{B}_1 \cap \mathcal{B}_2$. Then for any $x \in V$ there is a base element $\mathcal{B}_3$ containing $x$ and which is contained in $V$.

If $X$ is a topological space and $x \in X$, a neighbourhood basis at $x$ is a collection $\mathcal{B}_x$ of neighbourhoods of $x$ such that every neighbourhood of $x$ contains at least one $B \in \mathcal{B}_x$. A set is said to be countably infinite if it admits a bijection with the set of positive integers, and countable if it is finite or countably infinite. A topological space $X$ is said to be first-countable if there is a countable neighbourhood basis at each point and second-countable if there is a countable base for its topology. If $(X, \tau)$ is a topological space we call a subcollection $\mathcal{B}$ of $\tau$ a subbase for $\tau$ if $\mathcal{B}$ generates $\tau$. This means that $\tau$ is the smallest topology containing $\mathcal{B}$. Further we call a topological space $(X, \tau)$ compact if each of its open covers has a finite subcover, i.e., for every collection $\mathcal{C}$ of open subsets of $X$ with $X = \cup_{C \in \mathcal{C}} C$ there exists a finite subset $\mathcal{F}$ of $\mathcal{C}$ such that

$$X = \bigcup_{F \in \mathcal{F}} F \qquad (2.10)$$

Closed subsets of compact spaces are again compact and any finite union of compact spaces is compact. Thanks to the Tychonoff-Theorem, the product of any collection of compact spaces is compact. A sequence $(K_i)_{i=1}^{\infty}$ of compact subsets of a topological space $X$ is called an exhaustion of $X$ by compact sets if $X = \cup_i K_i$ and $K_i \subset \mathrm{int}\,(K_{i+1})$ for each $i$.

## 2.1.3  Locally Compact Spaces

**Definition 2.8**  Let $X$ be a topological space. We call $X$ locally compact, if every point of $x \in X$ has a compact neighbourhood. This means that there exists an open set $U$ and a compact set $K$ such that $x \in U \subset K$.

If $X$ is in addition a Hausdorff space, this property has two equivalent formulations that are often more useful. We call a subset of $X$ precompact in $X$ if its closure in $X$ is compact.

**Lemma 2.9**  For a Hausdorff space $X$, the following properties are equivalent

1)  $X$ is locally compact
2)  Each point of $X$ has a precompact neighbourhood
3)  $X$ has a basis of precompact open subsets

Now we are able to present properties of locally convex Hausdorff spaces, that are quite similar to properties of complete metric spaces. Hence this spaces can be handled easily and are therefore of interest.

**Theorem 2.10** Let $X$ be a locally compact Hausdorff space.

1) Every countable union of nowhere dense sets has empty interior
2) For $Y$ a topological space, every proper continuous map $f : Y \to X$ is closed
3) Every nonempty countable closed subset contains at least one isolated point
4) Is $X$ is addition second countable, then it admits an exhaustion by compact sets

## 2.1.4 Seminorms and Local Convexity

Often when vector spaces appear in the context of analysis, the notion of convergence of a sequence is defined with respect to a given norm. For instance, a popular example is the uniform convergence in $C[0, 1]$ by virtue of the supremum norm $\| \cdot \|_\infty$. However, within this framework the concept of pointwise convergence of a sequence of functions $(f_k)_{k \geq 1} \subset C[0, 1]$ can not be formulated. But in order to able to talk about convergence, a norm is far too much, i.e., a topological structure is already sufficient. Therefore, for $t \in [0, 1]$ and $(f_k)_{k \geq 1} \subset C[0, 1]$ define the map $p_t(f_k) := |f_k(t)|$, that is, the evaluation of the function $f_k$ at point $t$. The statement that the sequence of functions $(f)_{k \geq 1}$ converges pointwise to $f \equiv 0$, can then be stated as $p_t(f_k) \xrightarrow{k \to \infty} 0$ for all $t \in [0, 1]$.

**Definition 2.11** Let $X$ be a vector space. A seminorm $p : X \to \mathbb{R}$ is a function on $X$ such that

(a) $p(x + y) \leq p(x) + p(y)$ for all $x, y \in X$
(b) $p(\lambda x) = |\lambda| \, p(x)$ for all $\lambda \in \mathbb{F}$ and $x \in X$.

A family $\mathcal{P}$ of seminorms on $X$ is called separating if for each $x \neq 0$ there exists $p \in \mathcal{P}$ such that $p(x) \neq 0$. A convex set $A \subset X$ is called absorbing for $X$, if every $x \in X$ lies in $tX$ for some $t = t(x) > 0$. Further we call $A$ balanced, if $\lambda A \subset A$ for any $\lambda \in \mathbb{F}$ with $|\lambda| \leq 1$. The Minkowski functional $\mu_A$ of $A$ is defined for $x \in X$ via

$$\mu_A(x) := \inf \{t > 0 \, | \, t^{-1}x \in A\} \tag{2.11}$$

**Lemma 2.12** Suppose that $p$ is a seminorm on a vector space $X$. Then

(a) $|p(x) - p(y)| \leq p(x - y)$
(b) $\{x \in X \mid p(x) = 0\}$ is a linear subspace of $X$
(c) The set $B := \{x \in X \mid p(x) < 1\}$ is convex, balanced, absorbing and $p = \mu_B$

*Proof.* To (a). By the subadditivity of $p$ we have

$$p(x) = p(x + y - y) \leq p(x - y) + p(y) \Leftrightarrow p(x) - p(y) \leq p(x - y) \quad (2.12)$$

Since this also holds of the roles of $x, y$ are interchanged and using $p(x - y) = p(y - x)$, the claim follows. To (b). If $p(x) = p(y) = 0$ and $\lambda_1, \lambda_2 \in \mathbb{F}$, we non negativity of a seminorm implies $0 \leq p(\lambda_1 x + \lambda_2 y) \leq |\lambda_1| p(x) + |\lambda_2| p(y) = 0$. To (c). If $x \in B$, then also $\lambda x \in B$ for $|\lambda| \leq 1$ since $p(\lambda x) = |\lambda| p(x) \leq p(x) < 1$. Therefore $B$ is balanced. Further, if $x, y \in B$ and $t \in (0, 1)$, we have

$$p(tx + (1 - t)y) \leq tp(x) + (1 - t)p(y) < 1 \quad (2.13)$$

what yields the convexity of $B$. If $x \in X$ and $s > p(x)$, then $p(s^{-1}x) = s^{-1} p(x) < 1$, hence $B$ is absorbing and $\mu_B(x) \leq s$, implying $\mu_B \leq p$. But if $0 < t \leq p(x)$ then $p(t^{-1}x) \geq 1$ and thus $t^{-1}x \notin B$. Consequently $p(x) \leq \mu_B(x)$ what yields $p(x) = \mu_B(x)$. □

**Theorem 2.13** Let $X$ be a vector space and suppose $\mathcal{P}$ is a separating family of seminorms on $X$. For each $p \in \mathcal{P}$ and $n \in \mathbb{N}$ associate the set

$$U_{p,n} := \{x \in X \mid p(x) < \frac{1}{n}\} \quad (2.14)$$

Let $\mathcal{B}$ be the collection of all finite intersections of the sets $U_{p,n}$. Then $\mathcal{B}$ is a convex balanced local base for a topology $\mathcal{T}$ on $X$, which turns $X$ into a locally convex space such that the following properties hold

(a) every $p \in \mathcal{P}$ is continuous with respect to $\mathcal{T}$
(b) $E \subset X$ is bounded if and only if every $p \in \mathcal{P}$ is bounded on $E$.

*Proof.* Let $A \subset X$. We declare the set $A$ to be open if and only if a union of translates of members of $\mathcal{B}$. By construction this defined a translation-invariant topology $\mathcal{T}$ on $X$ as each member if $\mathcal{B}$ is convex and balanced and $\mathcal{B}$ is a local base for $\mathcal{T}$.

Suppose now that $x \in X \setminus \{0\}$. Since the family of seminorms is separating, there exists $p \in \mathcal{P}$ such that $p(x) \neq 0$. If $np(x) > 1$ then $x \notin U_{p,n}$ and therefore 0 is not contained in the neighbourhood $x \setminus U_{p,n}$ of $x$ and consequently $x$ is not in the closure of $\{0\}$. Thus $\{0\}$ is a closed set and by the translation invariance of $\mathcal{T}$, every set of the form $\{x\}$ with $x \in X$ is a closed set. Further we have to prove that the vector space operations addition and scalar multiplication are continuous with respect to $\mathcal{T}$. Let $N_0$ be a neighbourhood of 0 in $X$. Then there exist $p_1, ..., p_n \in \mathcal{P}$ and $n_1, ..., n_n \in \mathbb{N}$ such that

$$N \supset U_{p_1,n_1} \cap \cdots \cap U_{p_n,n_n} \tag{2.15}$$

Define $U := U_{p_1,n_1} \cap \cdots \cap U_{p_n,n_n}$. Since any seminorm $p \in \mathcal{P}$ is subadditive, also $U + U \subset N$, hence addition is continuous. For $x \in X$ and $\lambda \in \mathbb{F}$ let $U$ and $N$ as be constructed in (2.15). Then there exists $s > 0$ such that $x \in sU >$. Set $t := s(1 + |\lambda|s)^{-1}$. If $y \in x + tU$ and $|\kappa - \lambda| < s^{-1}$ we have $\kappa y - \lambda x = \kappa(x - y) + (\kappa - \lambda)x$ which lies in

$$|\kappa|tU + |\kappa - \lambda|sU \subset U + U \subset N \tag{2.16}$$

since $|\kappa|t \leq 1$ and $U$ is by assumption balanced. Consequently scalar multiplication is continuous. Summing up, $X$ is a locally convex space. Hence by Lemma 2.12 (a) the seminorm $p$ is continuous on $X$ with respect to $\mathcal{T}$. To show (b), suppose that $E$ is bounded and fix $p \in \mathcal{P}$. Since $U_{p,1}$ is a neighbourhood of 0, we have $E \subset kU_{p,1}$ for a sufficient large $k < \infty$. Therefore $p(x) < k$, for every $x \in E$ and every $p \in \mathcal{P}$ is bounded on $E$. Conversely assume that $E \subset X$ such that all $p \in \mathcal{P}$ are bounded on $E$. Let $U$ be a neighbourhood of 0 such that $U \supset U_{p_1,n_1} \cap ... \cap U_{p_n,n_n}$. Then there are numbers $M_i < \infty$ such that $p_i < M_i$ on $E$ for $i = 1, ..., n$. If $\ell > M_i n_i$ it follows $E \subset \ell U$, so that $E$ is bounded.                           $\square$

## 2.2   Vector Spaces With Order Unit

In the following we consider real vector spaces $V$ that are equipped with an additional structure. This structure is given by a cone $V^+ \subset V$, that contains the so called positive elements of $V$ and induces in a natural way a partial order. In the notation we are mainly following [26].

**Definition 2.14** Let $V$ be a real vector space. A subset $\emptyset \neq \mathcal{C} \subset V$ is called a cone in $V$ if the following properties are fulfilled

a)  $\lambda v \in C$ for all $\lambda \in [0, \infty)$ and $v \in C$
b)  $v + w \in C$ if $v, w \in C$
c)  $C \cap (-C) = \{0\}$

We say the cone $C$ is a full cone for $V$ if $C + (-C) = V$. Given points $x_1, \ldots, x_n \in V$ and $\alpha_1, \ldots, \alpha_n \in \mathbb{R}^+$, the point

$$x = \sum_{i=1}^{n} \alpha_i x_i \qquad (2.17)$$

is called a conic combination of the points $x_1, \ldots, x_n$. For a subset $S \subset V$ we denote by cone $(S)$ the set of all conic combinations of points from $S$ and call it the conic hull of the set $S$. The conic hull cone $(x)$ for $x \in V$ is called the ray spanned by $x$. For $C \subset V$ a cone and $\mathcal{K} \subset C$ a ray, we say that $\mathcal{K}$ is an extreme ray of $C$ if for any $u \in \mathcal{K}$ and any $x, y \in C$, whenever $u = (x + y)/2$, we must have $x, y \in \mathcal{K}$. Let $C$ be a cone and let $x \in C$ be a point. If $\mathcal{K} = $ cone $(x)$ is an extreme ray of $C$, we say that $x$ spans an extreme ray.

It is important to note that the definition of a cone differs in the literature. Thus property $a)$ is sometimes taken as definition of a cone. Is in addition property $b)$ fulfilled it is called a convex cone and $c)$ is called pointed. A function $f : K \to V$ from a convex set $K$ to a linear space $V$ is said to be affine if it preserves convex combinations, i.e., if

$$f (\lambda x + (1 - \lambda) y) = \lambda f (x) + (1 - \lambda) f (y) \qquad (2.18)$$

whenever $x, y \in K$ and $0 < \lambda < 1$. Thus, a real valued function on $K$ is affine if and only if it is both convex and concave.

**Definition 2.15**  Suppose $S$ is a set and that $\leq$ is a relation on $S$, i.e., a subset of the Cartesian product $S \times S$. Then $\leq$ is called a partial order if for all $x, y, z \in S$ the following properties are fulfilled

a)  $x \leq x$
b)  if $x \leq y$ and $y \leq x$ then $x = y$
c)  if $x \leq y$ and $y \leq z$ then $x \leq z$

A set equipped with a partial order is called a partial ordered set or poset. If $v \leq w$ implies $v + x \leq w + x$ for all $x \in V$ the partial order is called translation invariant and if $v \leq w$ implies $\alpha v \leq \alpha w$ for all $\alpha \in [0, \infty)$ one calls $\leq$ invariant under multiplication by non negative reals.

An ordered vector space is a tuple $(V, V^+)$ consisting of a real vector space $V$ and a cone $V^+ \subset V$. In this case, one also calls $V^+$ the positive cone of $V$ or the cone of positive elements, since $v \in V^+$ if and only if $0 \leq v$, thus $V^+$. A positive cone in an ordered vector space defines a partial order, but also the reverse is true. If an ordered vector space is given, it defines naturally a partial ordering on $V$ by $v \geq w$ if and only if $v - w \in V^+$. This partial ordering is translation invariant and invariant under multiplication of non negative reals. On the other hand given a partial ordering on $V$ that is translation invariant and invariant under multiplication with non negative reals, one can define the set $V^+ := \{v \in V \mid 0 \leq v\}$. With this the tuple $(V, V^+)$ becomes an ordered vector space. For the development of physical theories in the context of ordered vector spaces, called generalized probabilistic theories, the existence of an element that mimics a unit is important. This leads to the concept of an order unit. If $(V, V^+)$ is an ordered vector space, an element $e \in V$ is called an order unit for $V$ if for each $v \in V$ there exists a real number $r > 0$ such that $re \geq v$. Is in addition whenever $v \in V$ with $re + v \geq 0$ for all $r > 0$ already $v \in V^+$, then we call the order unit $e$ an Archimedean order unit. We call an ordered vector space with order unit an order unit space. Is in addition this order unit Archimedean, it is called Archimedean order unit space.

**Example 2.16** The existence of non Archimedean order units is not a peculiarity of infinite dimensional vector spaces but rather natural in finite dimensions. Consider the cone

$$V^+ = \left\{ (x, y) \mid x > 0 \text{ and } y \geq x \right\} \cup \left\{ (0, 0) \right\} \tag{2.19}$$

together with the order unit $e = (1, 2)$. Then $e$ is not an Archimedean order unit since $re + (0, 1) \in V^+$ for all $r > 0$ but $(0, 1) \notin V^+$.

**Lemma 2.17** ([26]). If $(V, V^+)$ is an ordered vector space with order unit $e$, then

a) $e \in V^+$
b) $V^+$ is a full cone of $V$

*Proof.* For the proof of $a$), note that by the definition of $e$ we can find $r > 0$ such that $re \geq -e$. The translation invariance of $\leq$ implies $(1 + r) e = re + e \in V^+$. Since $V^+$ is a cone and $\alpha := (1 + r)^{-1} > 0$ one has $e = \alpha (re + e) \in V^+$. To see $b$) let $v \in V$ and $r > 0$ such that $re \geq -v$ and $re \geq v$. With this one obtains $re + v \in V^+$ and $re - v \in V^+$ and thus $v = (re + v)/2 - (re - v)/2 \in V^+ - V^+$. Since $v \in V$ was arbitrary the claim follows.                                           □

**Definition 2.18**   A linear functional $\rho$ on an order unit space $A$ is called a state if it is positive on positive elements $v \in V^+$ and $\rho(e) = 1$. The set $S$ of all states of $A$ is called the state space of $A$. An extreme point of the state space is called a pure state.

In [26] an analogue of the Hahn-Banach theorem was proven which gives conditions under which one can extend a positive $\mathbb{R}$-linear functional on a subspace of $V$ to a positive $\mathbb{R}$-linear functional on the whole space $V$. If $(V, V^+)$ is an ordered vector space and $S \subset V$, we say that $S$ majorizes $V^+$ if for each $v \in V^+$ there is $w \in S$ such that $w \geq v$. Obviously, if $S$ majorizes $V^+$ and $S \subset T$, then also $T$ majorizes $V^+$. Further, if $(V, V^+)$ admits an order unit $e$ and $e$ is contained in a linear subspace $E$, then $E$ majorizes $V^+$. This can be seen as follows. By definition of an order unit, we can always find $r \geq 0$ such that $re \geq v$ for all $v \in V^+$. Since $e \in E$ we have $re \in E$ what shows the claim.

**Theorem 2.19**   ([26]). Let $(V, V^+)$ be an ordered vector space, such $V^+$ is a full cone for $V$. If $E \subset V$ is a subspace that majorizes $V^+$ and if $f : E \to \mathbb{R}$ is a positive $\mathbb{R}$-linear functional on $E$, then there exists a positive $\mathbb{R}$-linear functional $\tilde{f} : V \to \mathbb{R}$ such that $\tilde{f}|_E = f$.

An important consequence of this theorem is that the state space of $(V, V^+)$ is always nonempty. Order unit spaces have the property that one can introduce a seminorm in a natural way.

**Definition 2.20**   Let $(V, V^+)$ be an order unit space with order unit $e$. Define for $v \in V$

$$||v|| := \inf \{ r \in \mathbb{R} \mid re + v \geq 0 \ \text{ and } \ re - v \geq 0 \} \tag{2.20}$$

The map $|| \cdot ||$ is called the order seminorm on $V$ determined by $e$.

Clearly, the value of the map $|| \cdot ||$ depends on the particular choice of the order unit $e$. For instance, consider the ordered vector space introduced in Example 2.16 once with the order unit $e_1 = (1, 2)$ and once with $e_2 = (2, 3)$.

**Lemma 2.21** Let $(V, V^+)$ be an Archimedean order unit space with an order unit $e$. Then $V$ admits a norm given by

$$||v|| = \inf \{\lambda > 0 : -\lambda e \leq v \leq \lambda e\} \tag{2.21}$$

satisfying $-||v||e \leq v \leq ||v||e$.

*Proof.* For any $v \in A$ define

$$m(v) = \inf \{\alpha \in \mathbb{R} | v \leq \alpha e\} \quad l(v) = \sup \{\beta \in \mathbb{R} | \beta e \leq v\} \tag{2.22}$$

Clearly $m(\cdot)$ and $-l(\cdot)$ defining sublinear functionals. By definition we have $||v|| = \max \{m(v), l(v)\}$ and so $|| \cdot ||$ is a semi-norm. $|| \cdot ||$ is a valid norm if it fulfills $-||v||e \leq v \leq ||v||e$. We now show that this follows from Archimedicity. For all $\epsilon > 0$ we have that $v + \epsilon e \geq m(v)e$ and so $v - m(v) + \epsilon e \geq 0$ and since Archimedicity was assumed we have $v \geq m(v)e$. On the other hand we have $l(v) + \epsilon e \geq v$ and therefore $v \geq l(v)e$. Together we obtain $l(v)e \leq v \leq m(v)v$. $\qquad \square$

**Theorem 2.22** ([26]). Let $(A, A^+)$ be an order unit space with order unit $e$ and let $||\cdot||$ be the order seminorm on $V$ determined by $e$. Then the following are equivalent:

a) $e$ is Archimedean
b) $A^+$ is a closed subset of $A$ in the order topology induced by $|| \cdot ||$

**Theorem 2.23** ([29]). Let $(V, V^+)$ be an Archimedean order unit space with state space $S$. If $v \in V$, then

$$v \in V^+ \Leftrightarrow \rho(a) \geq 0 \; \forall \rho \in S \tag{2.23}$$

and

$$||v|| = \sup \{|\rho(v)| : \rho \in S\} \tag{2.24}$$

*Proof.* Clearly $a \in A^+$ implies that $\rho(a) \geq 0$ for all $\rho \in S$. To prove the other direction, let $\rho(a) \geq 0$ for all $\rho \in S$ and assume for contradiction $a \notin A^+$. Since $A^+$ is closed, one can find by a separation version of the Hahn-Banach theorem a $\varphi \in A^*$ and $\alpha \in \mathbb{R}$ such that $\varphi(a) < \alpha$ and $\varphi(b) \geq \alpha$ for all $b \in A^+$. Since $A^+$ is a cone, we can choose the separating number $\alpha$ to be zero. Hence $\varphi$ is a positive linear functional and $\rho := \varphi(e)^{-1} \varphi$ is a state with $\rho(a) < 0$. Contradiction. To prove the second claim define $\lambda = \sup\{|\rho(a)| : \rho \in S\}$. Obvious we have $\lambda \leq ||a||$. Assume for contradiction $\lambda < ||a||$. By definition of the order unit norm, either $a \notin (-\lambda e + A^+)$ or $a \notin (\lambda e - A^+)$. If $a \notin (-\lambda e + A^+)$ then $\lambda e + a \notin A^+$. In the above we have already proven that there exists $\rho \in S$ such that $\rho(\lambda e + a) < 0$. Hence $\rho(a) < -\lambda$, a contradiction. Similarly $a \notin (\lambda e - A^+)$ gives $\rho(a) > \lambda$ what is also contradiction. $\qquad\square$

**Lemma 2.24** Let $V, W$ be real vector spaces and $\mathcal{C} \subset V$ be a cone in $V$. Further let $\varphi : V \to W$ be a monomorphism.

a) Is $\mathcal{C}$ pointed, then $\varphi(\mathcal{C})$ is pointed
b) Is $\mathcal{K} \subset V$ an extreme ray in $\mathcal{C}$, then $\varphi(\mathcal{K})$ is an extreme ray in $\varphi(\mathcal{C})$

*Proof.* Since $\varphi$ is injective, we can write $\varphi(u)$ uniquely as $\varphi\left(\frac{x+y}{2}\right)$ and by linearity $\varphi(u) = \frac{1}{2}(\varphi(x) + \varphi(y))$. By assumption we have $x, y \in \mathcal{K}$ and hence $\frac{1}{2}(\varphi(x) + \varphi(y)) \in \varphi(\mathcal{K})$ since $\varphi(\mathcal{K})$ is a cone. Suppose $V^+ \cap (-V^+) = \{0\}$ and $\varphi(V^+) \cap \varphi(-V^+) = \text{cone}(\alpha)$ for $0 \neq \alpha \in W$. Since $\varphi$ is injective, there exists $\beta \in V^+ \wedge \beta \in -V^+$ such that $\varphi(\beta) = \alpha$ what is a contradiction to $V^+ \cap (-V^+) = \{0\}$. $\qquad\square$

## 2.3 Base-norm Spaces

Let $V$ be a real vector space and $V^+ \subset V$ a cone. A set $K \subset V^+$ is called a base for $V^+$ if $0 \notin K$ and for every point $0 \neq u \in V^+$ there exists a unique representation $u = \lambda v$ with $v \in K$ and $\lambda > 0$. Further we call the set $B := \text{conv}(K \cap -K)$ the symmetric convex hull of $K$. A set $B$ is said to be radially compact if the set $\{\lambda : \lambda x \in B\}$ is a compact subset of $\mathbb{R}$ for every $x \in B$ with $x \neq 0$. The notion of a radially compact set is strongly related to the concept of an absorbing set.

**Definition 2.25** An normed order unit space $V$ with a full cone $V^+$ is said to be a base norm space if the norm can be obtained by the Minkowski functional

$$||x|| = \inf \{\alpha > 0 \mid x \in \alpha B\} \qquad (2.25)$$

with $K$ base of $V^+$ such that $B = \operatorname{conv}(K \cup -K)$ is radially compact. The convex set $K$ is called the distinguished base of $V$.

**Lemma 2.26**  Let $V$ be a real vector space ordered by a full cone $V^+$. Assume that $V^+$ posses a basis $K$ located on a hyperplane $H \not\ni \{0\}$ and that its symmetric convex hull $B$ is radially compact. Then, the Minkowski functional induced by $B$ becomes a norm for $V$, i.e., $(V, || \cdot ||_B)$ turns into a normed space with the closed unit ball $B$. If $K$ is compact in some Hausdorff topology $\mathcal{J}$ compatible with the linear structure of $V$, then $V$ is complete in the norm (2.25).

*Proof.*  Clearly $B$ is a convex set and absorbing for $V$, thus by Theorem 2.2 the Minkowski functional is a seminorm. Since the set $B$ is also balanced it turns into a norm. If $K$ is a $\mathcal{J}$-compact set, then also $B$ and we can consider a Cauchy sequence $(x_n)$. Now $(x_n)$ is bounded and by a rescaling we get that $x_n \in B$ for $n \in \mathbb{N}$. Let $y$ be some $\mathcal{J}$-accumulation point of the sequence $(x_n)$ in $B$. We have to show that $y$ must be a norm limit of $(x_n)$. Since $(x_n)$ is a Cauchy sequence we can find $\epsilon > 0$ arbitrary and choose $n_0$ such that $||x_n - x_m|| \leq \epsilon$ for $n, m \geq n_0$. In particular we have $x_n \in x_{n_0} + \epsilon B$ for $n \geq n_0$. Since $\epsilon B$ is $\mathcal{J}$-closed we have that $y \in x_{n_0} + \epsilon B$, or $||y - x_{n_0}|| \leq \epsilon$. Thus we obtain the estimation

$$||y - x_n|| \leq ||y - x_{n_0}|| + ||x_{n_0} - x_n|| \leq 2\epsilon \qquad (2.26)$$

for all $n \geq n_0$.  $\square$

**Lemma 2.27**  If $V$ is a base norm space with distinguished base $K$, then the restriction map $f \mapsto f|_K$ is an order and norm preserving isomorphism of $V^*$ onto the space $\mathbb{A}_b(K)$ of all real valued bounded affine functions on $K$ equipped with pointwise ordering and supremum norm.

*Proof.*  For $f \in V^*$, the restriction to $K$ is a bounded affine function and the supremum of $|f(v)|$ over $B = \operatorname{conv}(K \cup -K)$ is the same as the supremum value over $K$, that is

$$||f|| = \sup_{v \in B} |f(v)| = \sup_{v \in K} |f(v)| \qquad (2.27)$$

On the other hand, every bounded affine function $f_0$ on $K$ can be uniquely extended to a bounded linear functional $f$ defined on all of $V$. In particular, for every $v \in V$

there exists $x, y \in K$ and $\alpha, \beta \in \mathbb{R}^+$ such that $v = \alpha x - \beta y$. Hence $f(v) = \alpha f_0(x) - \beta f_0(y)$ determines a well-defined linear functional on $V$.                                         □

**Theorem 2.28** ([29]). The dual of an order unit space $A$ is a base norm space and the dual of a base norm space $V$ is an order unit space. More specifically, the distinguished base of $A^*$ is the state space of $A$ and the distinguished order unit of $V^*$ is the unit functional on $V$.

*Proof.* Let $A$ be an order unit space. Thanks to Lemma 5.8 it is sufficient to show that the unit ball $B_1(A^*) := B_1^*$ of the Banach space $A^*$ is equal to the symmetric convex set conv $(K \cup -K)$ with $K$ the state space of $A$. The first inclusion conv $(K \cup -K) \subset B_1^*$ holds trivially. For the converse relation, assume that there is $\omega \in B_1^*$ such that $\omega \notin$ conv $(K \cup -K)$. Because $K$ is a w*-compact, the set conv $(K \cup -K)$ is as a subset w*-compact and consequently w*-closed. By the Hahn-Banach separation theorem there exists a w*-continuous linear functional on $A^*$ that separates the point $\omega$ from the w*-closed convex set conv $(K \cup -K)$. Since conv $(K \cup -K)$ is symmetric, the separating real number can be chosen to be $\alpha > 0$, in particular set $\alpha = 1$. Thus there exists $a \in A$ such that $\omega(a) > 1$ and $\sigma(a) < 1$ for all $\sigma \in$ conv $(K \cup -K)$. So $|\sigma(a)| \leq 1$ for all $\sigma \in K$, and so $\|a\| \leq 1$. Thus $|\omega(a)| \leq 1$ what is a contradiction. For the second claim let $K$ be the distinguished base of $V$. The space $\mathbb{A}_b(K)$ of bounded real valued affine functions on $K$ is an order unit space. By Lemma 2.27 $V^*$ maps isomorphically onto $\mathbb{A}_b(K)$, so also $V^*$ must be an order unit space. The linear functional $e$ is constantly 1 on $K$, so it maps to the unit functional 1 in $\mathbb{A}(K)$. Hence $e$ is the distinguished order unit in $V^*$.                                         □

Here it is unit space and its dual have quite different structures. While the order unit space is an ordered space with a preferred element in its positive cone, the dual is a base norm space, i.e., an ordered space with a preferred base $\Omega$ for the positive cone. Indeed, the space $(V, V^+, e_V)$ and its dual $(V^*, \Omega)$ are generally not isomorphic as ordered spaces. If there exists a linear order isomorphism between this spaces, we say that $(V, V^+, e_V)$ is weakly self-dual. Provided that this isomorphism induces an inner product on $(V, V^+, e_V)$ such that

$$V^+ = \{v \in V \mid \langle v, w \rangle \geq 0 \, \forall w \in V\} \tag{2.28}$$

we call $(V, V^+, e_V)$ self-dual. Further we call $(V, V^+, e_V)$ homogeneous if the group of order-automorphisms $\alpha : V \to V$ acts transitively on the interior of $V^+$. Recall that a group $G$ acts transitively on a set $X$ if for any pair $x, y \in X$ there exists a $g \in G$ such that $g \cdot x = y$. Clearly if the group act transitively, the space of coinvariants $X_G$, i.e., the set of orbits of $X$ under the group $G$, is a singleton.

## 2.4 Functional Representation

Similar to the Gelfand representation for commutative $C^*$-algebras, we want a representation of Archimedean order unit spaces as continuous real valued functions on a compact Hausdorff space $X$. To formulate the problem more generally we are looking for fruitful procedures to investigate abstract algebraic structures. Here we are following the approach of [30] and using the functional representation method. For an introduction to locally convex Hausdorff spaces we refer to [66]. First we have to fix some notation.

**Definition 2.29** If $K$ and $K'$ are convex subsets of a locally convex Hausdorff space $E$ over $\mathbb{R}$ and $K \subset K'$, then $\mathbb{A}(K, K')$ denotes the vector space of all restrictions to $K$ of continuous affine real-valued functions on $K'$. We write $\mathbb{A}(K)$ in place of $\mathbb{A}(K, K)$. For X locally compact Hausdorff space, we write

$$C_{\mathbb{R}}(X) = \{f : X \to \mathbb{R} \mid f \text{ is continuous}\} \qquad (2.29)$$

In general the following inclusion holds true

$$\mathbb{A}(K, E) \subset \mathbb{A}(K, K') \subset \mathbb{A}(K) \qquad (2.30)$$

Also note that if $K$ is compact $\mathbb{A}(K)$ is a uniformly closed subspace of $C_{\mathbb{R}}(K)$.

**Lemma 2.30** ([31]). Let $\psi : (A, e) \to (A', e')$ be a linear map between Archimedean order unit spaces with order unit $e$ and $e'$ respectively, such that $\psi(e) = e'$. Then $\psi$ is order preserving or positive if and only if $\psi$ is bounded with $||\psi|| = 1$. Is in addition $\psi$ bijective, $\psi$ is an order isomorphism, i.e., $\psi$ and $\psi^{-1}$ are positive, if and only if $\psi$ is an isometry.

*Proof.* Since we have to show an equivalence, assume first that $\psi$ is positive and let $a \in A$ with $||a|| \leq 1$. It follows that $-e \leq a \leq e$ and by positivity $\psi(-e) = -e' \leq \psi(a) \leq e' = \psi(e)$. Hence $||\psi(a)|| \leq 1$ and so $||\psi|| \leq 1$. But $e$ and $e'$ are order units what implies $||\psi(e)|| = ||e'|| = 1$ and $||e|| = 1$ what shows $||\psi|| = 1$. To prove the other direction, assume that $||\psi|| = 1$ and consider a positive element $a \in A$. W.l.o.g. assume $||a|| \leq 1$, i.e., $0 \leq a \leq e$. Now $0 \leq e - a \leq e - a + a = e$ and so $||e - a|| \leq ||e|| = 1$. Since $\psi$ was by assumption contractive, we have

$||\psi\,(e-a)\,|| \leq 1$ and so $\psi\,(e-a) \leq e'$. Hence $\psi\,(a) \geq 0$. Is $\psi$ now in addition bijective, the necessity is obvious. Sufficiency follows from the above, since an isometry has the property $||\psi|| = ||\psi^{-1}|| = 1$.                                   $\square$

A subset $F$ of a compact Hausdorff space $X$ is said to be a max-boundary for a cone $\mathcal{K} \subset C_{\mathbb{R}}\,(X)$ if for each $f \in \mathcal{K}$ there exists a point $y \in F$ such that $f\,(y) = \sup_{x \in X} f\,(x)$. Let $X$ be a compact Hausdorff space and $(A, e)$ an order unit space. An isomorphismus $\rho\,:\,(A, e) \rightarrow (C_{\mathbb{R}}\,(X), 1_X)$ will be called a functional representation of $(A, e)$ over $X$ and we denote this representation by $(\rho, X)$. A functional representation is said to be separating if $\rho\,(A)$ separates the points of $X$. For $(\rho, X)$ and $(\sigma, Y)$ two functional representations of $(A, e)$ we say that $(\rho, X)$ is larger than $(\sigma, Y)$ if there exists a homeomorphism $\varphi\,:\,Y \rightarrow X$ such that $\sigma = \varphi^* \circ \rho$, in particular $(\sigma a)\,(y) = (\rho a)\,(\varphi y)$ for all $a \in A$ and $y \in Y$. If $(\rho, X)$ is a functional representation of $(A, e)$ and if a closed subset $Y$ of $X$ is a max-boundary for $\rho\,(A)$ in that $||a|| = ||\rho a|| = \sup_{y \in Y} |\,(\rho a)\,(y)\,|$ or all $a \in A$, then the canonical injection $\varphi\,:\,Y \rightarrow X$ induces a functional representation $(\varphi^* \circ \rho, Y)$ which is called the restriction of $(\rho, X)$ to $Y$. As a consequence of Lemma 2.30 we obtain that every order unit space $(A, e)$ admits a functional representation. One of those is the canonical representation $(\rho, S)$ over the state space, defined by $A \ni a \mapsto \hat{a}$ where $\hat{a}\,(p) = p\,(a)$ for all $p \in S$.

**Theorem 2.31**  (Kadison [31]). The canonical representation $(\rho, S)$ of an order unit space $(A, e)$ over its state space $S$ is the largest separating functional representation of $(A, e)$. Its range $\rho\,(A)$ consists of all those w*-continuous functions on $S$ which can be extended to w*-continuous linear functionals on the surrounding space $A^*$. In particular, $\rho\,(A) = A\,(S)$ if and only if $(A, e)$ is complete in the order unit norm. The restriction of $(\rho, S)$ to $\overline{\partial_e S}$ is the smallest separating functional representation of $(A, e)$.

As already mentioned above, this theorem is quite similar to the Gelfand representation for commutative $C^*$ algebras. This may indicate that this is part of a more general procedure. Indeed, both are part of the class of theorems describing topological algebraic structures and are therefore called "characterization theorems" for $C\,(X)$ where $X$ is a compact Hausdorff space. In more detail, Theorem 2.31 provides a general representation for partially ordered vector spaces. This representation can be used to characterize linear subspaces of $C\,(X)$ containing the constant function. From this result one can derive representation theorems for a different kind of topologized structures e.g. the algebra theorem of Stone for ordered algebras [32], the Stone-Krein-Kakutani-Yosida lattice theorem [33, 34], the characterization of $C\,(X)$ as a Banach space [35–37] and Banach algebra representations [38, 39].

## 2.5 Archimedeanization and Categories

In this section we give a short introduction how one can build tensor products of order unit spaces such that the algebraic properties are preserved. Given two real or complex vector spaces $V$, $W$ there are at least two well known operations under which the vector space structure should behave well. First if $U \subset V$ is a linear subspace, then also the quotient $V / U$ should be a vector space. Further the tensor product $V \otimes W$ should lead again to a vector space. As we will see, this problem is closely related to the Archimedeanization of order unit spaces and category theory. In particular, the category of order unit spaces and unital positive maps misbehaves under functorial operations such as quotients and tensor products.

**Definition 2.32** Let $(V, V^+)$ be an order unit space with order unit $e$. Then a subspace $I \subset V$ is called an order ideal provided that $c \in I$ and $0 \leq b \leq c$ implies that $b \in I$.

**Lemma 2.33** If $I$ is an order ideal in the order unit space $(V, V^+)$ with order unit $e$ then the quotient space $(V / I, V^+ / I)$ is an order unit space with order unit $e + I$.

*Proof.* Clearly $V^+ + I$ is a cone and $e + I$ serves as an order unit. It remains to show that $(V^+ + I) \cap (-V^+ + I) = \{0 + I\}$. For this assume that there exists $v, w \in V^+$ such that $v + I = -w + I$. This implies the existence of $u \in I$ such that $v = -w + u$ and hence $u = v + w \in V^+$. By definition we have $0 \leq v \leq u$ such that $v \in I$. Thus $v + I = 0 + I$. $\qquad\square$

The difficulties in the present situation stem from the possibility of factoring an Archimedean order unit space $V$ by a closed ideal and getting an order unit space which is not Archimedean. Thus we are interested in a process that assign to every order unit space an Archimedean order unit space. To formalize this concept mathematically we need the notion of a category and a functor.

**Definition 2.34** A category $C$ consists of

a) a class of objects denoted by $\mathrm{Obj}\,(C)$ or just $C$
b) a class of morphisms or arrows between this objects denoted by $\mathrm{Hom}\,(C)$
c) a law of composition for morphisms, i.e., for every three objects $A$, $B$, $C$ a binary operation $\mathrm{Hom}\,(A, B) \times \mathrm{Hom}\,(B, C) \to \mathrm{Hom}\,(A, C)$

such that the following axioms hold:

1) associativity, i.e., for $A, B, C, D \in C$, $f \in \mathrm{Hom}\,(A, B)$, $g \in \mathrm{Hom}\,(B, C)$ and $h \in \mathrm{Hom}\,(C, D)$ one has $h \circ (g \circ f) = (h \circ g) \circ f$
2) existence of an identity, i.e., for every object $A \in C$ one has a unique morphism $1_A \in \mathrm{Hom}\,(A, A)$ called identity morphism of $A$ such that $f 1_A = f$ for every $f \in \mathrm{Hom}\,(A, B)$ and $1_A f = f$ for every $f \in \mathrm{Hom}\,(B, A)$.

A category $C$ is called a subcategory of a category $D$ if every object of $C$ is also an object of $D$ and if $A, B \in C$, then $\mathrm{Hom}\,(A, B)$ in $C$ is contained in $\mathrm{Hom}\,(A, B)$ in $D$. Let $C$ and $D$ be categories. A functor $\mathcal{F}$ from $C$ to $D$ is a mapping that

a) associates to each object $X$ in $C$ an object $\mathcal{F}(X)$ in $D$
b) associates to each morphism $f : X \to Y$ in $C$ a morphism $\mathcal{F}(f) : \mathcal{F}(X) \to \mathcal{F}(Y)$ in $D$ such that

1) $\mathcal{F}(1_X) = 1_{\mathcal{F}(X)}$ for each object $X \in C$
2) $\mathcal{F}(g \circ h) = \mathcal{F}(g) \circ \mathcal{F}(h)$ for all $f \in \mathrm{Hom}\,(X, Y)$ and $g \in \mathrm{Hom}\,(Y, Z)$

For $(V, V^+)$ an order unit space with order unit $e$ let $D = \{v \in V \,|\, re + v \in V^+ \,\forall r > 0\}$ and $N = D \cap -D$. We define $V_{Arch}$ to be the order unit space $(V / N, (V / N)^+)$ with the Archimedean order unit $e + N$. We call $V_{Arch}$ the Archimedeanization of $V$. The following result describes a universal property that characterizes the Archimedeanization

**Theorem 2.35** ([26]). Let $(V, V^+)$ be an order unit space with order unit $e$ and let $V_{Arch}$ be the Archimedeanization of $V$. Then there exists a unital surjective positive linear map $q : V \to V_{Arch}$ with the property that whenever $(W, W^+)$ is an order unit space with Archimedean order unit $e'$ and $\varphi : V \to W$ is a unital positive linear map, then there exists a unique positive linear map $\tilde{\varphi} : V_{Arch} \to W$ with $\varphi = \tilde{\varphi} \circ q$. In terms of commutative diagrams we have

$$V \xrightarrow{\ q\ } V_{Arch}$$
$$\underset{\varphi}{\searrow} \quad \downarrow{\tilde{\varphi}}$$
$$W$$

In addition, this property characterizes $V_{Arch}$, i.e., if $V'$ is another ordered vector space with Archimedean order unit and $q' : V \to V'$ is a unital surjective positive linear map with the above property, then $V' \cong V_{Arch}$.

It turns out [40] that the order unit spaces together with unital positive maps forming in a natural way a category $\mathcal{O}$. Further the Archimedean order unit spaces together with unital positive maps forming a subcategory $\mathcal{O}_{Arch}$ of $\mathcal{O}$. With Theorem 2.35 the Archimedeanization can be viewed as a functor $\mathcal{F}$ from the category $\mathcal{O}$ to the subcategory $\mathcal{O}_{Arch}$. Further this functor fixes the subcategory $\mathcal{O}_{Arch}$, i.e., it is a projection onto this subcategory. Let $(V, V^+)$ be an Archimedean order unit space with order unit $e$ and let $I \subset V$ be an order ideal. Further let $(V / I, V^+ + I)$ be the quotient of $V$ by $I$. Then we call the Archimedeanization of $(V / I, V^+ + I)$ the Archimedean quotient of $V$ by $I$.

## 2.6   Tensor Products

In this section we introduce the notion of tensor products in order unit spaces. For Hilbert spaces or nuclear spaces there is a simple well-behaved theory of tensor products. If one is interested in such a theory based on Banach spaces or locally convex topological vector spaces the theory is notoriously subtle. However, it turns out that there is a huge variety of possibilities to define a tensor product. To be more precise, there is a continuum of possibilities. For the application on a physical theory only the "extreme" cases are needed. From an algebraic perspective, one is especially interested in their functorial properties. Before we can give a formal definition what we mean by a tensor product of function systems, we need a precise notion of the tensor product for vector spaces.

### 2.6.1   Tensor Products of Vector Spaces

Let $V$, $W$ be vector spaces over the field $\mathbb{F} = \{\mathbb{R}, \mathbb{C}\}$. A tensor product of $V$ and $W$ is given by a vector space $Z$ and a bilinear map $\tau : V \times W \to Z$ with the following universal property: For any $\mathbb{F}$- bilinear map $\psi : V \times W \to E$ into a vector space $E$ there exists a unique map $\varphi : Z \to E$ with $\psi = \varphi \circ \tau$ such that the diagram

$$V \times W \xrightarrow{\ \tau\ } Z$$
$$\searrow^{\psi} \quad \downarrow^{\varphi}$$
$$E$$

commutes. Since tensor products are defined via a universal property, they are unique up to canonical isomorphisms. Instead of $\tau\,(V \times W)$ we usually write $V \otimes W$ and for $x \in V$ and $y \in W$ we write for $\tau\,(x, y) = x \otimes y$ and call these objects tensors in $V \otimes W$. By using this notation, we can write the tensor product in the for physicist more convenient form

$$V \times W \to V \otimes W \quad , \quad (x, y) \mapsto x \otimes y \tag{2.31}$$

## 2.6.2  Tensor Products of Order Unit Spaces

Because order unit spaces can be used to characterise the space $C\,(X)$ for some locally compact Hausdorff space $X$, they are also called function systems. Especially if one deals with tensor products it is more convenient to use this notation. Suppose $V, W$ are function systems and let $\varphi : V \to C\,(X)$ and $\Psi : W \to C\,(Y)$ be faithful representations with $X, Y$ locally compact Hausdorff spaces. This leads to the tensor product of representations defined by $\varphi \otimes \Psi : V \otimes W \to C\,(X \times Y)$ what is again a faithful representation. We define the minimal tensor product $V \otimes_{\min} W$ as the function system structure on $V \otimes W$ induced by $\varphi \otimes \Psi$. It can be proven that the minimal tensor product is independent from the choice of faithful representations $\varphi, \Psi$ [41]. Apart from this abstract characterization of the minimal tensor product, it turns out that there is an alternative description. The positive cone of the minimal tensor product $V \otimes_{\min} W$ can be written as

$$(V \otimes_{\min} W)^{+} = \left\{ z \in V \otimes W : (f \otimes g)\,(z) \geq 0 \ , \ f \in S\,(V) \ , \ g \in S\,(W) \right\} \tag{2.32}$$

Unfortunately, the minimal tensor product of two Archimedean order unit spaces is in general not again an Archimedean order unit space, but still an order unit space. By applying the Archimedeanization procedure one obtains an Archimedean order unit space. Clearly it would be desirable to have an operation that preserves the Archimedean property, i.e., a map $\otimes_{\min}\,(V \times W) = V \otimes_{\min} W$ such that $V \otimes_{\min} W$ is an order unit space and coincides with its Archimedeanization. In fact, the maximal tensor product possesses this property.

**Definition 2.36**  Let $V$, $W$ be function systems and denote with $(V \otimes W)^*$ the algebraic dual of $V \otimes W$. We define the maximal state space as

$$S_{max}(V \otimes W) := \left\{ \rho \in (V \otimes W)^* : \rho|_{V^+ \otimes W^+} \geq 0 \right\} \tag{2.33}$$

with

$$V^+ \otimes W^+ := \left\{ \sum_{i=1}^{n} v_i \otimes w_i : n \in \mathbb{N}, \ v_i \in V^+, \ w_i \in W^+ \right\} \tag{2.34}$$

The maximal tensor product is defined as the function system structure on $V \otimes W$ induced by the inclusion $V \otimes W \subset C(S_{max}(V \otimes W))$.

For $v \in V$ and $w \in W$ we have $||v|| \, ||w|| e_V \otimes e_W \pm v \otimes w = \frac{1}{2}(||v|| e_v \pm v) \otimes (||w|| e_w + w) + \frac{1}{2}(||v|| e_v \mp v) \otimes (||w|| e_w - w)$. By definition of the order norm we have $||v|| e_v \pm v \in V^+$ and thus $||v|| \, ||w|| e_V \otimes e_W \pm v \otimes w \in V^+ \otimes W^+$. This proves that $e_V \otimes e_W$ is a valid order unit. Since $V \otimes W$ is again a real vector space and $V^+ \otimes W^+$ a cone, we conclude that $(V \otimes W, V^+ \otimes W^+, e_V \otimes e_W)$ is an order unit space.

**Theorem 2.37**  ([40]). Suppose that $(V, V^+, e_V)$ and $(W, W^+, e_W)$ are function systems. Then the maximal tensor product $V \otimes_{max} W$ coincides with the Archimedeanization $(V \otimes W, D, e_V \otimes e_W)$ of $(V \otimes W, V^+ \otimes W^+, e_V \otimes e_W)$.

*Proof.* First show "$\subset$". Let $z \in (V \otimes_{max} W)^+$ and $f$ a state on $(V \otimes W, D, e_V \otimes e_W)$. Clearly one has $V^+ \otimes W^+ \subset D$ and hence $f|_{V^+ \otimes W^+} \geq 0$ so $f \in S_{max}(V \otimes W)$. It follows that $f(z) \geq 0$ for any state on $(v \otimes W, D, e_V \otimes e_W)$ and thus $z \in D$. To show "$\supset$" assume that $z \in D$ and $f \in S_{max}(V \otimes W)$. Since $z + \epsilon e_V \otimes e_W \in V^+ \otimes W^+$ for $\epsilon > 0$, we have

$$0 \leq f(z + \epsilon e_V \otimes e_W) = f(z) + \epsilon \tag{2.35}$$

This implies that $f(z) \geq 0$ and therefore $z \in (v \otimes_{max} W)^+$.  □

The Kronecker product $\otimes_K \equiv \otimes$ which is used in quantum mechanics for the formulation of joint systems is neither the minimal nor the maximal one. In fact it turns out to be in between with respect to inclusion, i.e.,

$$(M_m (\mathbb{R}) \otimes_{\max} M_n (\mathbb{R}))^+ \subset (M_m (\mathbb{R}) \otimes M_n (\mathbb{R}))^+ \subset (M_m (\mathbb{R}) \otimes_{\min} M_n (\mathbb{R}))^+$$
$$(2.36)$$

Here the question arises under which conditions the minimal and the maximal tensor product of two Archimedean order unit spaces coincide. In analogy to notation in the theory of $C^*$-algebras, we say that a pair $(V^+, W^+)$ is nuclear, if $V \otimes_{\max} W = V \otimes_{\min} W$. This question dates back to the work of Barker [42] in the 70s. There it was proven that a pair $(V^+, W^+)$ is nuclear whenever either $V^+$ or $W^+$ is isomorphic to the $\mathbb{R}^d_+$ with $d$ the dimension of the cone. Further it was conjectured [42] that also the converse holds. This question was recently answered in [43].

## 2.7   $C^*$-algebras

**Definition 2.38**   A Banach algebra is a Banach space $(A, ||\cdot||)$ that is in addition an algebra in which

$$||ab|| \leq ||a|| \, ||b|| \quad \forall a, b \in A \tag{2.37}$$

An involution on an algebra $A$ is a $\mathbb{R}$-linear map $* : A \to A$ with $a \mapsto a^*$ such that $a^{**} = a$, $(ab)^* = b^* a^*$ and $(\lambda a)^* = \bar{\lambda} a^*$ for all $a, b \in A$ and $\lambda \in \mathbb{C}$. An algebra with an involution is called a $*$-algebra. A $C^*$-algebra is a Banach algebra $A$ with an involution in which

$$||a^* a|| = ||a||^2 \quad \forall a \in A \tag{2.38}$$

**Example 2.39**   There are two main examples of $C^*$-algebras that will appear frequently in this thesis.

(1) Let $X$ be a locally compact Hausdorff space and denote with $C_0(X)$ the set of all continuous functions $f : X \to \mathbb{C}$ that vanish at infinity. If one equip this set with pointwise operations i.e., $(fg)(x) = f(x)g(x)$ and $(\lambda f + g)(x) = \lambda f(x) + g(x)$ this turns $C_0(X)$ into an algebra. Furthermore there is a natural involution inherited from $\mathbb{C}$, namely $f^*(x) = \overline{f(x)}$ and a natural norm $||f||_\infty = \sup_{x \in X} |f(x)|$. The algebra $C_0(X)$ is unital if and only if $X$ is compact. In this case the identity element is given by the function $f(x) = 1$ for all $x \in X$. The most important property of $(C_0(X), ||\cdot||_\infty)$ is that it constitutes a commutative $C^*$-algebra.

(2) Let $\mathcal{H}$ be a Hilbert space and $B(\mathcal{H})$ the set of all bounded linear operators from $\mathcal{H}$ to itself, with the obvious algebraic operations $(+, \cdot)$ and the involution given by the adjoint. The norm is taken to be the operator norm, i.e., for $x \in B(\mathcal{H})$ we have $\|x\| = \sup\{\|x\xi\| \mid \xi \in \mathcal{H}, \|\xi\| = 1\}$. It is a unital $C^*$-algebra, where the unit is given by the identity operator $1_{\mathcal{H}}$. If $\dim(\mathcal{H}) > 1$, $B(\mathcal{H})$ is a non-commutative $C^*$-algebra.

To be more precise, Definition 2.38 is the definition of a abstract $C^*$-algebra, which is strongly motivated the structure of $B(\mathcal{H})$, as introduced in Example 2.39. Moreover, each operator norm-closed $*$-algebra in $B(\mathcal{H})$ is a $C^*$-algebra. As we will see in the following, this are in fact all possible examples.

**Definition 2.40**   A representation of a $C^*$-algebra $A$ is a $*$-homomorphism from $A$ to $B(\mathcal{H})$ for some Hilbert space $\mathcal{H}$. Two representations $\pi$ and $\rho$ of $A$ on Hilbert spaces $X$ and $Y$ respectively are unitarily equivalent if there is a unitary operator $U \in B(X, Y)$ with $U\pi(x)U^* = \rho(x)$ for all $x \in A$. A subrepresentation of a representation $\pi$ on $\mathcal{H}$ is the restriction of $\pi$ to a closed invariant subspace of $\mathcal{H}$. A representation is irreducible if it has no closed invariant subspaces. If $I \neq \emptyset$ index set and $\pi_i$ a representation of $A$ on $\mathcal{H}_i$, then the sum $\oplus_i \pi_i$ of the $\pi_i$ is the diagonal sum acting on $\oplus_i \mathcal{H}_i$. If each $\pi_i$ is equivalent to a fixed representation $\rho$, then $\oplus_i \pi_i$ is called amplification of $\rho$ by $|I|$. A representation with kernel 0 is called faithful. We call a representation $\pi$ cyclic, if its carrier space $\mathcal{H}$ contains a cyclic vector $\xi$ for $\pi$, i.e., the closure of $\pi(A)\xi$ coincides with $\mathcal{H}$ and it is called nondegenerate that $\pi(a)v = 0$ for all $a \in A$ and $v \in \mathcal{H}$ implies $v = 0$.

**Theorem 2.41**   Each $C^*$-algebra $A$ is isomorphic to a norm-closed $*$-algebra in $B(\mathcal{H})$, for some Hilbert space $\mathcal{H}$.

The proof of Theorem 2.41 relies on the so called GNS-construction. This *ingenious* construction discovered independently by Gelfand and Naimark [2, 3] and I. Segal [8] is one of the most fundamental ideas of the theory of operator algebras and provides a method for manufacturing representations of $C^*$-algebras.

**Theorem 2.42**   (GNS construction) Let $A$ be a unital $C^*$-algebra [1] and $\phi$ a state on $A$. There exists a cyclic representation $\pi_\phi$ of $A$ on a Hilbert space $\mathcal{H}_\phi$ with cyclic unit vector $\xi_\varphi$ such that

---

[1] The theorem remains true even if $A$ is not unital.

$$\phi(a) = \langle \xi_\phi, \pi_\phi(a)\xi_\phi \rangle \quad a \in A \tag{2.39}$$

*Proof.* We first consider the special case where $\phi(a^*a) > 0$ for all $a \in A \setminus \{0\}$. Define a sesquilinear from $(\cdot, \cdot)$ on $A$ by $(a, b) := \phi(a^*b)$. Since $\phi$ is a state, the sesquilinear form is positive and we can complete $A$ in the ensuing norm given by $\|a\|_\phi = \sqrt{\phi(a^*a)}$ to a Hilbert space $\mathcal{H}_\phi$. For each $a \in A$ define the map $\pi_\phi(a) : A \to A$ via $\pi_\phi(a)b = ab$, i.e., the left multiplication by $a \in A$. Regarding $A$ as a dense subspace of $\mathcal{H}_\phi$ this defines an operator $\pi_\phi(a)$ on a dense domain in $\mathcal{H}_\phi$. This operator is bounded since $\|\pi_\phi(a)\| \leq \|a\|$. Hence we can extend $\pi_\phi(a)$ from $A$ to $\mathcal{H}_\phi$ by continuity and we obtain a map $\pi_\phi : A \to B(\mathcal{H}_\phi)$. A direct calculation shows that $\pi_\phi$ is indeed a representation. The cyclic vector $\xi_\phi$ is given by the unit $1 \in A$, seen a an element of $\mathcal{H}_\phi$. Clearly, this element is cyclic and we have

$$\|\xi_\phi\|^2 = \langle \xi_\phi, \xi_\phi \rangle = \phi(1^*1) = \phi(1) = 1 \text{ and } \langle \xi_\phi, \pi_\phi(a)\xi_\phi \rangle = \phi(1^*a1) = \phi(a) \tag{2.40}$$

Suppose now that there exists states $\phi$ on $A$ such that $\phi(a^*a) = 0$ for some $a \in A$. For an arbitrary state $\phi$ define

$$N_\phi := \{a \in A \mid \phi(a^*a) = 0\} \tag{2.41}$$

and consider the quotient space $A/N_\phi$. If $[a]_\phi$ denotes the image of $a \in A$ under the projection in $A/N_\phi$ we can define an inner product on $A/N_\phi$ via $\langle [a]_\phi, [b]_\phi \rangle := \phi(a^*b)$. This form is well defined and positive definite, hence we can define the Hilbert space $\mathcal{H}_\phi$ as the completion of $A/N_\phi$ if this inner product. Furthermore define

$$\pi_\phi(a) : A/N_\phi \to \mathcal{H}_\phi \text{ with } [b]_\phi \mapsto \pi_\phi(a)[b]_\phi := [ab]_\phi \tag{2.42}$$

This map $\pi_\phi(a)$ is well defined for each $a \in A$, since $N_\phi$ is a left ideal in $A$. If we define in addition $\xi_\phi = [1]_\phi$, the claim follows.                    $\square$

## 2.8   Completely Positive Maps

We denote by $M_n(\mathbb{C})$ the set of all $n \times n$ matrices with complex coefficients. Note that one can identify the space $M_m(\mathbb{C}) \otimes M_n(\mathbb{C})$ with the space $M_n(M_m(\mathbb{C}))$ the collection of all $n \times n$ block matrices with $m \times m$ matrices as entries. This identification can be easily seen if one considers the map $\iota : M_m (\mathbb{C}) \otimes M_n (\mathbb{C}) \to M_n (M_m (\mathbb{C}))$ with $\iota (A \otimes B) = \left( b_{ij} A \right)_{1 \leq i, j \leq m}$. It is simple to check that $\iota$ is a linear and bijective map, thus an isomorphism.

**Definition 2.43** A linear map $\phi : M_n (\mathbb{C}) \to M_m (\mathbb{C})$ is called positive if $\phi (A)$ positive for all positive $A \in M_n (\mathbb{C})$. Define the map

$$\phi \otimes \mathbb{I}_p : M_n \otimes M_p \to M_m \otimes M_p \tag{2.43}$$

$\phi$ is called completely positive if $\phi \otimes \mathbb{I}_p$ is positive for all $p \in \mathbb{N}$.

Since the matrix algebra $M_n (\mathbb{C})$ is just a special representative of a $C^*$-algebra, we formulate the following theorems in this slightly more general language. In particular, in the language of the $C^*$-algebra of bounded operators on a Hilbert space. Since the Gelfand-Naimark theorem allows to identify an arbitrary $C^*$-algebra isometrically $*$-isomorphic with a $C^*$-algebra of bounded operators on a Hilbert space, this are in principle all $C^*$-algebras. Further it is important to note, that for $\mathcal{H}, \mathcal{K}$ Hilbert spaces any unital $*$-homomorphism $\pi : \mathcal{L} (\mathcal{H}) \to \mathcal{L} (\mathcal{K})$ is also completely positive. Namely, if $A \in \mathcal{L} (\mathcal{H})$ is positive, one can find $B \in \mathcal{L} (\mathcal{H})$ such that $A = B^* B$. Then

$$\pi (A) = \pi \left( B^* B \right) = \pi (B)^* \pi (B) \tag{2.44}$$

hence $\pi$ is positive. Since $\pi$ is a unital $*$-homomorphism it follows that also the induced map $\pi \otimes \mathbb{I}_p$ is a unital $*$-homomorphism and thus $\pi$ is completely positive.

**Theorem 2.44** ([44]). Let $\mathcal{A}$ be a unital $C^*$-algebra and let $\phi : \mathcal{A} \to \mathcal{L} (\mathcal{H})$ be a completely positive map with $\mathcal{H}$ Hilbert space. Then there exists a Hilbert space $\mathcal{K}$, a unital $*$-homomorphism $\pi : \mathcal{A} \to \mathcal{L} (\mathcal{K})$ and a bounded operator $V : \mathcal{H} \to \mathcal{K}$ with $|| \phi (1) || = ||V||^2$ such that

$$\phi (a) = V^* \pi (a) V \tag{2.45}$$

Note that any map of the form $\phi (a) = V^* \pi (a) V$ is completely positive. Therefore the Stinespring dilation Theorem 2.44 characterizes the completely positive maps from any $C^*$-algebra into the algebra of bounded operators on any Hilbert space. Further if the map $\phi$ is unital, then the bounded operator $V$ is an isometry. We will call the tripel $(\pi, V, \mathcal{K})$ as in Theorem 2.44 a Stinespring representation for $\phi$.

Let $\mathcal{H}$ be a separable Hilbert space and $\{e_i\}_{i \in \mathbb{N}}$ an orthonormal basis. We call a $T \in \mathcal{L}(\mathcal{H})$ a trace class operator if

$$\mathrm{tr}\,(|T|) = \sum_{i \in \mathbb{N}} \langle e_i | T e_i \rangle < \infty \tag{2.46}$$

with $|T| := \sqrt{T^*T}$. We denote by $\mathcal{I}(\mathcal{H})$ the set of trace class operators. It can be easily seen, that $\mathcal{I}(\mathcal{H})$ is not only a linear subspace of $\mathcal{L}(\mathcal{H})$ it is an ideal. Further we call a linear map $\phi : \mathcal{I}(\mathcal{H}) \to \mathcal{I}(\mathcal{H})$ a channel if it is completely positive and trace preserving.

**Corollary 2.45** A linear mapping $\phi : \mathcal{I}(\mathcal{H}) \to \mathcal{I}(\mathcal{H})$ is channel if and only if there exists a sequence of bounded operators $(A_k)_{k \in \mathbb{N}}$ such that

$$\phi(T) = \sum_{k \in \mathbb{N}} A_k T A_k^* \quad , \quad \sum_k A_k^* A_k = \mathbb{I} \tag{2.47}$$

The form (2.47) is called an operator-sum form, or Kraus form of the channel $\phi$. The operators $A_k$ are also called Kraus operators.

## 2.9   Convex Polytopes

**Definition 2.46** Let $K$ be a convex subset of $\mathbb{R}^d$. A set $F \subset K$ is a face of $K$ if either $F = \emptyset$ or $F = K$, or if there exists a supporting hyperplane $H$ of $K$ such that $F = H \cap K$. The set of all faces of $K$ is denoted by $\mathfrak{F}(K)$. A point $x \in K$ is an exposed point of $K$ if the set $\{x\}$ consisting of the single point $x$ is a face of $K$. The set of all exposed points is denoted by $\exp(K)$. A compact convex set $K \subset \mathbb{R}^d$ is called a polytope if $\exp(K)$ is a finite set. For a polytope $K$ we call the points of $\exp(K)$ vertices and denote their totality by $\mathrm{vert}(K)$. 1-faces are called edges, while maximal proper faces are facets of $K$. Further we write $f_k(K)$ for the number of different $k$-faces of a polytope $K$. A set $K \subset \mathbb{R}^d$ is called a polyhedral set provided $K$ is the intersection of a finite family of closed halfspaces of $\mathbb{R}^d$.

**Theorem 2.47** (Caratheodory [45]). If $A$ is a subset of $\mathbb{R}^d$ then every $x \in \mathrm{conv}(A)$ is expressible in the form

$$x = \sum_{i=0}^{d} \alpha_i x_i \quad \text{where} \quad x_i \in A, \ \alpha_i \geq 0 \text{ and } \sum_{i=0}^{d} \alpha_i = 1 \qquad (2.48)$$

In Chapter 4 we will often use the property that any polytope can we written as an intersection of a finite family of halfspaces. Also it shows that polytopes are polyhedral sets. The following theorem formalises this fact.

**Theorem 2.48** ([45]). Each $d$-polytope $K \subset \mathbb{R}^d$ is the intersection of a finite family of closed halfspaces. The smallest such family consists of those closed halfspaces containing $K$ whose boundaries are the affine hulls of the facets of $K$.

*Proof.* Denote by $\mathcal{H} := \{H_j \mid 1 \leq j \leq f_{d-1}(K)\}$ the set of hyperplanes determined by the facets of $K$ and let a point $y \notin K$ be given. We have to show that there exists a $H_j$ such that $y$ does not belong to the closed halfspace determined by $H_j$ and containing $K$. We denote by $L$ the set of all affine combinations of at most $d - 1$ points of vert $(K)$. By Theorem 2.47 the set $L$ contains all the faces of $K$ which have dimension at most $d - 2$. Let $M$ denote the cone spanned by $L$ with vertex $y$. Then $M$ is contained in the union of finitely many hyperplanes through $y$. Since finitely many hyperplanes do not cover any nonempty open set, int $(K)$ is not contained in $M$. Let $x$ be any point of $(\text{int}(K))^c$ and consider the ray $N := \{\lambda x + (1 - \lambda) y \mid \lambda \geq 0\}$ with endpoint $y$ determined by $x$. It follows that $N \cap \text{int}(K) \neq \emptyset$. We write $\lambda_0 = \inf\{\lambda > 0 \mid \lambda x + (1 - \lambda) y \in K\}$. Since $K$ is compact and $y \notin K$, the greatest lower bound is attained, $0 < \lambda_0 < 1$ and $x_0 = \lambda_0 x + (1 - \lambda_0) y \in \partial K$. Thus $x_0$ belongs to some proper face $F$ of $K$. But $x \notin M$ implies $x_0 \notin L$ and therefore $F$ is not of dimension less than or equal to $d - 2$. Thus $F$ is a facet and the hyperplane aff $(F)$, the set of all affine combinations, has all the desired properties. The assertion about the minimality of $\mathcal{H}$ is obvious, which completes the proof. $\quad\square$

# Generalized probabilistic theories 3

## 3.1 Preparation and Measurements

As it is well known, the standard formulation of quantum mechanics involves concepts as global complex phases, that are not directly accessible by our sensory experience. The framework of generalized probabilistic theories (GPTs) tries to avoid such concepts as much as possible by defining a theory operationally in terms of preparation procedures, transformations and measurements. A typical experimental setup in physics consists of a preparation procedure, possibly followed by a sequence of manipulations or transformations, and a final measurement. Obviously ambiguities can occur whether the intermediate manipulations are part of the preparation procedure or the measurement. Hence the experimental setup can be further abstracted to preparations and measurements only. It is important to note, that in standard quantum mechanics the absorption of the intermediate transformation into the preparation corresponds to the Schrödinger picture, i.e., the states evolve in time $|\psi\rangle = |\psi(t)\rangle$. The form of this transformation is fixed by the Schrödinger equation and the corresponding Hamiltonian. On the other hand, the absorption of the manipulations into the measurement process match to the Heisenberg picture, i.e., observables evolve within time governed by the Heisenberg equation. A measurement apparatus yields classical results and thus GPTs are exclusively concerned with the classical probabilities of measurement outcomes for a given preparation procedure. Because any measurement device, digital as well as analog, has only a finite resolution one can restrict to the case of finite many possible outcomes. For an individual measurement apparatus one can associate with each possible outcome a characteristic one-bit quantity which gives 1 if the result occurred and 0 otherwise. In this way a measurement can be decomposed into mutually excluding classical bits. The other way around, every single measurement can be interpreted as an appli-

J. Steinberg, *Extensions and Restrictions of Generalized Probabilistic Theories*, BestMasters, https://doi.org/10.1007/978-3-658-37581-2_3

cation of a collection of such 1-bit measurements. Clearly this procedure becomes more involved if one deals with several different measurement devices where the classical bits are not necessarily mutually excluding.

One of the basic assumptions of the GPT framework is that experiments can be repeated under identical conditions in such a way that the outcomes are statistically independent. Suppose one has access to a toolbox containing a finite number of one-bit measurements labeled by $m = 1, ..., M$ and a finite number of preparation procedures labeled by $k = 1, ..., K$. By construction each pair $(m = i, k = j)$ produces an outcome $\alpha\,(i, j) := \alpha_{ij} \in \{0, 1\}$. Repeating the experiment the specific outcome $\alpha_{ij}$ is typically not reproducible, indeed one can reproducibly estimate the probability $p_{ij} = \langle \alpha_{ij} \rangle$. For a given toolbox the values of $p_{ij}$ can be listed in a probability table. To remove redundancies in this table, we have to introduce the notion of operational equivalence, that is strongly related to the concepts of states and effects. Two preparation procedures are called operationally equivalent if it is impossible to distinguish them experimentally, i.e., any available measurement device responds of both with the same probability. In the same manner one calls two one-bit measurements operationally equivalent if both respond with the same probability to any of the available preparation procedures.

**Definition 3.1** ([4, 46]) A state $\omega$ is a class of operationally equivalent preparation procedures. An effect $e$ is a class of operationally equivalent 1-bit measurements. We denote by $e\,(\omega) = p\,(e|\omega)$ the probability that an experiment chosen from the equivalence classes $e$ and $\omega$ gives a positive outcome.

If one raise the claim that GPTs can describe realistic experiments, one has to deal with noise. Noise can be interpreted as a kind of classical randomness that can be taken into account by the notion of mixed states and effects. This mixtures should affect the probability for a certain outcome in the usual way. Suppose one has a preparation procedure that produces the state $\omega_1$ with probability $p$ and another procedure that produces the state $\omega_2$ with probability $1 - p$. Likewise the experimenter selects the effect $e_1$ with probability $q$ and effect $e_2$ with probability $1 - p$. The probability in the end to get a positive outcome is given by

$$P = p\,[qe_1\,(\omega_1) + (1 - p)\,e_2\,(\omega_1)] + (1 - p)\,[qe_1\,(\omega_2) + (1 - p)\,e_2\,(\omega_2)] \tag{3.1}$$

If one treats this more conceptual, one can impose that probabilistic mixtures of preparation procedures can be considered as a preparation procedure itself, meaning it defines a new state $\omega$. In particular $\omega = p\omega_1 + (1 - p)\omega_2$ and

$e = qe_1 + (1 - q) e_2$. Note that the concept of mixtures is not a priori clear and one can build theories without probabilistic mixtures [47]. We call this new objects mixed states and mixed effects. As pure states cannot be interpreted as statistical mixtures of other states, they are also called states of maximal knowledge. Furthermore, there is no physical distinction between states that can be prepared exactly, and states that can be prepared to arbitrary accuracy. Thus, we assume that the state space is topologically closed.

In reality experiments are not only noisy they also can be unreliable, i.e., they sometimes fails to produce a result. For instance a preparation procedure fails to create a certain object or a detector fails to detect a particle. If our preparation procedure creates a physical state with certainty or the measurement device responds to a particle with certainty it is called reliable. If a given effect $e$ responds to a specific state $\omega$ with certainty $e(\omega) = 1$ both are reliable. If $e(\omega) < 1$ there is no way to decide whether the reduced probability is caused by an unreliable state, an unreliable effect or by the probabilistic structure of the theory. To avoid this problem, we assume that there exists a special reliable effect, which checks whether a preparation was successful or not. This normalization functional $u$ attains the value one on all normalized states or reliable states. The normalization $u(\omega)$ can be interpreted as the probability of success in the preparation procedure. For states with $u(\omega) < 1$ the preparation succeeds with probability $u(\omega)$. At this point we are able to make our definition of a measurement more precise. A measurement with $n$ outcomes is a collection of effects $(e_1, ..., e_n)$ such that $e_1 + ... + e_n = u$. A set of states $\omega_1, ..., \omega_n$ is called perfectly indistinguishable if there exists a measurement such that $e_i(e_j) = \delta_{ij}$.

So far, we only have a description of a single system. However most of the striking phenomena of quantum mechanics emerge for composite systems, e.g., entanglement, cloning, broadcasting or teleportation protocols. Apart from their appearance in the setting of quantum mechanics, there is another reason why one is interested in this kind of phenomena in the convex framework. This is related to a classical result of Vinberg [48] and Koecher [49].

**Theorem 3.2** ([48, 49]) Let $V$ be a finite dimensional order unit space with cone $V^+$. If the cone is in addition self-dual, homogeneous and irreducible, i.e., the cone is not the direct sum of simpler cones, then the cone is either the cone of positive self-adjoint elements of some full matrix $*$-algebra over the reals, complexes or quaternions, or the cone generated by a ball-shaped base, or is the set of positive self-adjoint $3 \times 3$ matrices over the octonions.

Consequently self-duality together with irreducibility and homogeneity leads to a structure very much like Hilbert space quantum mechanics. The declared aim is now to motivate these conditions in operation terms, i.e., in terms of cloning, broadcasting, teleportation or multipath interference. In fact [50], one can identify weak self-duality as a necessary condition for probabilistic teleportation and a condition similar to homogeneity as sufficient for the existence of deterministic teleportation. To make this clear, we will shortly reformulate the definition of tensor products of order unit spaces for the state space. Note that while in mathematics the order unit space is taken to be fundamental, in physics it is convenient to take the state space $\Omega$ as the fundamental object. As in Chapter 2, there is no unique way, rather a spectrum of candidates, bounded by a maximal and a minimal tensor product. Since the state space $\Omega$ is by definition given by by a convex set, Theorem 2.28 asserts that $\mathbb{A}_b(\Omega)$, i.e., the set of all bounded real valued affine functions on $\Omega$, is an Archimedean order unit space. Thus one can introduce in analogy to Definition 2.36 a tensor product $\Omega \otimes \Omega'$ of two state spaces $\Omega$ and $\Omega'$ as the set of all bilinear functionals $\mu : \mathbb{A}_b(\Omega) \times \mathbb{A}_b\left(\Omega'\right) \to \mathbb{R}$. The maximal tensor product $\Omega \otimes_{\max} \Omega'$ is then given by all $\mu$ that are positive on pairs $(a, b)$ with $a, b \geq 0$ and normalized, i.e., $\mu\left(e, e'\right) = 1$ where $e, e'$ the order units of $\mathbb{A}_b(\Omega)$ and $\mathbb{A}_b\left(\Omega'\right)$ respectively. The minimal tensor product $\Omega \otimes_{\min} \Omega'$ of $\Omega$ and $\Omega'$ is the convex hull of the set of product states in $\Omega \otimes \Omega'$. We term such a convex combination a separable state. The set $\Omega \otimes_{\max} \Omega' \setminus \Omega \otimes_{\min} \Omega'$ contains the entangled states. If the dimensions of the vector spaces $\mathbb{A}_b(\Omega)$ and $\mathbb{A}_b\left(\Omega'\right)$ are finite it follows that they are isomorphic to their double duals and thus $\Omega \otimes_{\max} \Omega$ and $\Omega \otimes_{\min} \Omega$ have the same affine dimension. Hence every state in $\Omega \otimes_{\max} \Omega'$ can be written as an affine combination, i.e.,

$$\Omega \otimes_{\max} \Omega \ni \omega = \sum t_i \, \alpha_i \otimes \beta_i \ , \quad \sum t_i = 1 \tag{3.2}$$

In the following the connection between the tensor product of state spaces and the tensor product of Archimedean order unit spaces is established.

**Lemma 3.3** Let $A$ and $B$ be ordered vector spaces. If the dimensions of both are finite we have

$$(A \otimes_{\max} B)^* = A^* \otimes_{\min} B^* \quad \text{and} \quad (A \otimes_{\min} B)^* = A^* \otimes_{\max} B^* \tag{3.3}$$

*Proof.* In finite dimensions it is sufficient to prove only the first statement. By assumption we have that $\dim(A), \dim(B) < \infty$ and thus $\dim(A \otimes B) = \dim(A) \cdot \dim(B) < \infty$. Further the canonical map $\Phi : (A \otimes B) \to (A \otimes B)^{**}$ with

$a \otimes b \mapsto \Phi (a \otimes b)$ where $\Phi (a \otimes b) : (A \otimes B)^* \to \mathbb{R}$ with $\varphi \mapsto \varphi (a \otimes b)$ is an isomorphism. Suppose that we have already proven the first statement. Then

$$\left( A^* \otimes_{\max} B^* \right)^{**} = \left( A^{**} \otimes_{\min} B^{**} \right)^* \iff (A \otimes_{\min} B)^* = \left( A^* \otimes_{\max} B^* \right) \quad (3.4)$$

what is the second claim. Note that in the finite dimensional case one has $(A \otimes B)^* \cong A^* \otimes B^*$ as vector spaces. Therefore it remains only to show the equality of the positive cones. Now let $\gamma \in (A \otimes_{\max} B)^*$, i.e., $\gamma$ is a map of the form $\gamma : A \otimes_{\max} B \to \mathbb{R}$ such that $\gamma (\omega) \geq 0$ if and only if $\omega (a, b) \geq 0$ for all positive functionals $a \in A^*$ and $b \in B^*$. In particular, $\gamma$ is positive on all positive linear combinations of pure tensors. Hence $\gamma \in A^* \otimes_{\min} B^*$. On the other hand we identify the algebraic tensor product $A^* \otimes B^*$ with the space of all bilinear forms on $A^{**} \times B^{**}$, i.e., if $\alpha \in A^*$ and $\beta \in B^*$ we identify the pure tensor $\alpha \otimes \beta$ with the bilinear form $(\alpha \otimes \beta) (a, b) = a (\alpha) b (\beta)$ with $a \in A^{**}$ and $b \in B^{**}$. By definition of the minimal tensor product we have $\gamma = \sum_i p_i \alpha_i \otimes \beta_i$ with $p_i \geq 0$. Clearly $\gamma$ is then positive on all positive forms $\omega \in A \otimes B$ and thus an element of $(A \otimes_{\max} B)^*$. $\quad \square$

To obtain a full physical theory it remains to declare what one means by a physical process or equivalently what kind of dynamics is admissible within the theory. Following [50] we present processes by positive, norm contractive, linear mappings that have as input an initial systems with state space $A$ and as output a final system with state space $B$. In mathematical terms this means that a process is given by $\phi : A \to B$ with $||\phi|| \leq 1$. We interpret $||\phi (\alpha)|| = u (\phi (\alpha))$ as the probability that the process when the input is $\alpha$. Equivalently one can say that the effect $u \circ \phi \in A^*$ recording the occurrence of the process [51, 52]. It is convenient to impose more structure on the possible dynamics of a system represented by an abstract state space. For this scope we want that the dynamic is given by a dynamical semigroup, i.e., a closed convex set $\mathfrak{D}_A$ of the norm contractive positive linear maps $\tau : A \to A$ closed under composition and containing the identity map $\mathbb{I}_A$. A state space equipped with a dynamical semigroup is called a dynamical model. Let $\tau_A \in \mathfrak{D}_A$ be a physical process on the state space of system $A$ and $\tau_B \in \mathfrak{D}_B$ be a physical process on the state space of system $B$. It is a naturally assumption that then for any state $\omega$ on the composite system $AB$ also

$$(\tau_A \otimes \tau_B) (\omega) : a, b \mapsto \omega \left( \tau_A^* a, \tau_B^* b \right) \quad (3.5)$$

is a state on $AB$. We call the composite system $AB$ dynamically admissible, if (3.5) holds for any state. It is an important observation that elements of the tensor product $A \otimes B$ as well as elements of its dual $(A \otimes B)^*$ can be regarded as operators $A^* \to B$

and $A \to B^*$. This is due to the fact that we identify the algebraic tensor product $A \otimes B$ of two vector spaces $A$, $B$ with the space of bilinear forms on $A^* \times B^*$. Indeed, every $f \in (A \otimes B)^*$ induces $\hat{f} : A \to B^*$ with $\hat{f}(\alpha)(\beta) = f(\alpha \otimes \beta)$. Similar any $\omega \in A \otimes B$ induces a linear map $\hat{\omega} : A^* \to B$ in virtue of $\hat{\omega}(f)(g) = (f \otimes g)(\omega)$.

## 3.2    Attributes of GPTs

GPTs were initially invented in order to deepen our understanding of quantum theory and to figure out the properties, which make quantum theory special. From an information theoretic viewpoint, the quantum no-cloning theorem, the quantum no-broadcasting theorem and quantum teleportation protocols are features that distinguish quantum theory from its classical counterpart. As we will see in the following, most of this features are not special from a GPT perspective in the sense, that in most of the probabilistic theories the results can be reproduced.

**Cloning**
A deterministic cloning procedure for a state $\alpha \in \Omega$ involves preparing the system in a state $\alpha$, preparing a second copy of the system in a particular state $\beta$ and perform an operation on the combined system $\Omega \otimes \Omega$ such that the initial state $\alpha \otimes \beta$ maps to the final state $\alpha \otimes \alpha$. One can prove, that in GPTs cloning is generally impossible at least in finite dimensional non classical theories [53]. Clearly in the GPT framework, the existence of a cloning procedure will depend not only on the structure of the convex sets of states but also on what kind of affine mappings one admits as physical operations, i.e., procedures one can implement. If the state space just consist of one element $\Omega = \{\omega\}$, the map $\omega \mapsto \omega \otimes \omega$ is affine and thus there exists a cloning procedure. Thus in any theory where affine mappings between state spaces are physically realizable, every state is cloneable if one not demands that some map clones more than this one state.

**Definition 3.4** ([54]). A finite collection $\alpha_1, ..., \alpha_n$ of states is called

1) co-cloneable if there exists a single cloning map $\kappa : \Omega \to \Omega \otimes \Omega$ that clones them all, i.e., $\kappa(\alpha_i) = \alpha_i \otimes \alpha_i$ for all $i = 1, ..., n$
2) jointly distinguishable if there exists an observable or POVM $E = (a_0, ..., a_n)$ with $\alpha_j(a_i) = \delta_{ij}$. In this case we say that the $\alpha_i$ are distinguished by $E$.

In finite dimensional quantum theory the pure states corresponding to vectors $v$, $w$ in a Hilbert space are distinguishable if and only if the vectors are orthogonal. For general quantum states $\omega_1$, $\omega_2$, pure or mixed, it turns out that they are distinguishable if and only if the corresponding density operators $\omega_1 \omega_2 = \omega_2 \omega_1 = 0$.

**Theorem 3.5** ([54]). In any finite dimensional probabilistic theory, using any tensor product, distinct states are co-cloneable if and only if they are jointly distinguishable.

**Broadcasting**

Broadcasting can be seen as a generalization of cloning. To give a rigorous definition of broadcasting in the convex framework, we first need the notion of marginals.

**Definition 3.6** Let $\mathcal{B} : \Omega \to \Omega \otimes \Omega$ be an affine mapping. The marginals are given by the mappings $\mathcal{B}_1, \mathcal{B}_2 : \Omega \to \Omega$ with

$$\mathcal{B}_1 (\omega) (a) = \mathcal{B} (\omega) (a \otimes u) \text{ and } \mathcal{B}_2 (\omega) (b) = \mathcal{B} (\omega) (u \otimes b) \qquad (3.6)$$

where $u$ is the unit functional. This fact allows us to define conditional states $\omega_{2,a}$ and $\omega_{1,b}$ by

$$\omega_{2,a} (b) := \frac{\omega (a, b)}{\omega_1 (a)} \text{ and } \omega_{1,b} (a) := \frac{\omega (a, b)}{\omega_2 (b)} \qquad (3.7)$$

what lead to the expected identities $\omega (a, b) = \omega_1 (a) \omega_{2,a} (b) = \omega_{1,b} (a) \omega_2 (b)$. We say that $\omega \in \Omega$ is broadcast by $\mathcal{B}$ if $\mathcal{B}_1 (\omega) = \mathcal{B}_2 (\omega) = \omega$, that is if $\omega$ is simultaneously a fixed point of both $\mathcal{B}_1$ and $\mathcal{B}_2$.

If we denote by $\Gamma$ the set of all states that one can broadcast by $\mathcal{B}$, it is clear that $\Gamma$ is a convex set. Indeed, it is $\Omega$-affine, i.e., it is the intersection of $\Omega$ with an affine subspace. As already mentioned above, one can see cloning as a special case of broadcasting e.g. for pure states of $\Omega$, broadcasting reduces to cloning. Suppose that $\omega \in \Omega$ is extreme and $\mathcal{B} (\omega)$ has marginals equal to $\omega$. Further let $E = (a_1, ..., a_n)$ be an observable. Then we have

$$\omega_2 = \sum_{a \in E} \omega_1 (a) \omega_{2,a} \qquad (3.8)$$

This gives $\omega_2$ as a convex combination of the states $\omega_{2,a}$ with coefficients $\omega_1 (a)$. As $\omega_2$ is pure, we have for each element $a \in E$ either $\omega_1 (a) = 0$ or $\omega_{2,a} = \omega_2$. In either case $\omega (a, b) = \omega_1 (a) \omega_2 (b)$ for all effects $b$. Since $E$ was chosen arbitrary, this holds also for all effects $a$. Thus $\omega = \omega_1 \otimes \omega_2$.

**Theorem 3.7** ([54]). Let $\Gamma$ be the set of states broadcast by an affine mapping $\mathcal{B} : \Omega \to \Omega \otimes \Omega$. Then $\Gamma$ is a simplex generated by jointly distinguishable states in $\Omega$.

## Teleportation protocols

As we have seen above, no-cloning theorems and no-broadcasting theorems are quite generic features of any non-classical probabilistic theory and not specifically quantum at all. For teleportation protocols the situation changes. In fact, most of such theories do not allow for teleportation. Suppose $ABC$ is a composite system of state spaces $A$, $B$ and $C$. If $f$ is an effect on $AB$ and $\omega$ state on $BC$ they induce the positive mappings $\hat{f} : A \to B^*$ and $\hat{\omega} : B^* \to C$ such that their composite $\hat{\omega} \circ \hat{f}$ is a positive operator $A \to C$. Further if $\alpha \in A$ and $c \in C^*$ any effect, we have that $\alpha \otimes \omega$ is a state in $ABC$ and $f \otimes c$ is an effect in $(ABC)^*$.

**Definition 3.8** ([50]). A pair $(f, \omega)$ is called a conclusive teleportation protocol on $A_1 A_2 B$ if there exists a norm contractive linear mapping $\tau : B \to B$, called a correction, such that, for every normalized state $\alpha \in \Omega_{A_1}$

$$\tau \left( \left( \alpha \, \tilde{\otimes} \, \omega \right)_f^B \right) = t_\alpha \eta \left( \alpha \right) \tag{3.9}$$

for some constant $t_\alpha > 0$. Here $\tilde{\otimes}$ refers to the corresponding normalized conditional state. If $\tau$ can be chosen that $t_\alpha = 1$ for all $\alpha$, we say that the protocol $(f, \omega)$ is strong. Is in addition $B \cong A_1$ via a fixed isomorphism $\eta : A_1 \to B$ and is $\omega$ a state in $A_2 B$ with $E = (f_1, ..., f_n)$ an observable on $A = A_1 A_2$, we say that the pair $(E, \omega)$ realizes a deterministic teleportation protocol if for each effect $f_i \in E$, the pair $(f_i, \omega)$ realizes s strong conclusive teleportation protocol.

Let $A$ be an abstract state space and consider a finite group $G$ acting on $A$ such that the state space is preserved. This induces a dual acting of $G$ on $A^*$ in virtue of

$$(ga)(\alpha) = a \left( g^{-1} \alpha \right) \quad \forall g \in G \; , \; a \in A^* \; , \; \alpha \in A \tag{3.10}$$

**Theorem 3.9** ([50]). Let $A$ be weakly self-dual and suppose that $G$ is a finite group acting on $A$ in such a way that $(i)$ $G$ acts transitively on the extreme points of $\Omega$ and $(ii)$ there exists a $G$-equivariant isomorphism $A^* \cong A$. Then $A \otimes_{min} (A \otimes_{max} A)$ supports a deterministic teleportation protocol.

## 3.3 Examples of GPTs

**Classical probability theory**

A set of discrete classical events defines an order unit lattice. It is the $n$-fold Cartesian product of $(\mathbb{R}, \mathbb{R}^+, 1)$, where $n$ is the number of outcomes. The set of states is given by maps

$$\mathbf{v} \mapsto \langle \mathbf{p}, \mathbf{v} \rangle \text{ with } \mathbf{p}_k \geq 0 \text{ and } \sum_{k=1}^{n} \mathbf{p}_k = 1 \tag{3.11}$$

Therefore the state space $\Omega$ forms a simplex, i.e., a probability simplex where the extreme states $\omega_i$ form a basis for dual space. This leads to a unique decomposition of mixed states into the pure ones. In particular the state space is given by a Choquet-simplex and thus leads to a nuclear function system. That means that $\Omega_A \otimes_{\max} \Omega_B = \Omega_A \otimes_{\min} \Omega_B$. Consequently there are no entangled states. Further the extreme effects $e_i$ are connected to the extreme states $\omega_j$ via $e_i(\omega_j) \propto \delta_{ij}$ and the system is in a particular pure state all the time. Mixed states are only an effective representation due to a lack of knowledge.

**Quantum mechanics**

An important example of a GPT is that of quantum mechanics itself. We choose the bounded self-adjoint operators as vector space $V$ and we identify $V^+$ to be the set of positive semi-definite operators. With the choice $e = \mathbb{I}$ this forms an Archimedean order unit space. To fix notation, a $n$-level quantum system is described by an order unit space where $V \subset M_n(\mathbb{C})$ is the set of self-adjoint matrices. The effects are then given by $F \in V$ with $0 \leq F \leq \mathbb{I}$. The state space $\Omega$ is given by

$$\{\omega \in V^* \mid \omega(F) = \text{tr}(\rho F) : \rho \in V^+, \text{tr}(\rho) = 1\} \tag{3.12}$$

The quantum mechanical tensor cone $V_{AB}^+$ is given by the set of positive semidefinite operators in $M_{n_A}(\mathbb{C}) \otimes M_{n_B}(\mathbb{C})$. The state space $S_{AB}^{\min}$ corresponding to $(V_A \otimes_{\min} V_B)^+$ is precisely the set of separable states and the state space $S_{AB}^{\max}$ corresponding to $(V_A \otimes_{\max} V_B)^+$ is given by the set of all positive semidefinite operators $W$ with $\text{tr}(W) = 1$ which satisfy $\text{tr}(WA \otimes B) \geq 0$ for all $A$ and $B \geq 0$. This set is dual to the set of entanglement witnesses [55] and includes all legal density operators as well as some operators with negative eigenvalues. Apart from the fact that quantum mechanics allows entanglement there are several other properties which are from the GPT perspective special, namely teleportation and broadcasting protocols. If there is no entanglement between two parties, Alice and Bob, but shared randomness, they can produce local correlations. Local correlations can always be written in the form

$$p_{ab|xy} = \sum_i p_i \, p_{a|xi} \, p_{b|yi} \qquad (3.13)$$

Quantum correlations are generated if two parties share quantum entanglement. The conditional probabilities can be calculated by the formula

$$p_{ab|xy} = \text{tr}\left[\rho\left(M_{a|x} \otimes M_{b|y}\right)\right] \qquad (3.14)$$

where Alice's and Bob's measurements are denoted by $M_{a|x}$ and $M_{b|y}$ respectively. Note that $x$ and $y$ denote the choice of measurement and $a$ and $b$ the corresponding outputs. The set of quantum correlation is convex but not a polytope.

**Popescu-Rohrlich box**
So far, we have just discussed well known theories where the framework of GPTs is not necessary to apply. In the following we discuss an example that is neither classical nor quantum and the language of order unit spaces is appropriate. It is well known that quantum mechanics is a nonlocal theory. But is quantum mechanics the unique theory that allows for nonlocal phenomena consistent with special relativity? Surprisingly the answer to this question is no. It turns out that nature could be even more nonlocal then quantum mechanics predicts, yet be fully consistent with relativity. When Bell discovered nonlocality, the problem was not formulated in a model independent way, i.e., it relies on the formalism of self-adjoint operators, eigenvalues and entangled quantum states. One possibility to get rid of this dependence is to view experiments as input-output black-boxes. In this formalism Alice has a black box that accepts input $x$ and yields outputs $a$, where the input $x$ indicates which experiment inside the box is to be performed. Thus the entire physics is encapsulated in the distribution $p\,(a|x)$, i.e., the probability that the measurement $x$ yields outcome $a$. Since we are interested in correlations, we consider in the following a second party, called Bob, who has also a black box available. Further we assume that Alice and Bob are situated far away from each other, i.e., their experiments are space-like separated. The physics here is encapsulated in $p\,(a, b|x, y)$, the joint probability that Alice obtains output $a$ and Bob output $b$ when Alice inputs $x$ and Bob inputs $y$. Popescu and Rohrlich [56, 57] found an assignment to the probabilities $p\,(a, b|x, y)$ which fulfill the non-signalling constraint, i.e.,

$$p(a|x) = \sum_b p(a, b|x, y) = \sum_b p(a, b|x, \tilde{y}) \quad \forall y, \tilde{y} \qquad (3.15)$$

and the same for $p(b|y)$, while reaching the algebraic maximum of the CHSH-inequality. Hence boxes that are in agreement with non-signalling but reaching the algebraic maximum of the CHSH-inequality are called PR-boxes. Here we restrict to a simple example of a PR-box (Table 3.1), where the probability assignment is given by

$$p_{ab|xy} = \begin{cases} \frac{1}{2} & a+b \bmod 2 = xy \\ 0 & \text{otherwise} \end{cases} \qquad (3.16)$$

This leads to the following probability table

**Table 3.1** Probability distribution of a PR-box

| y \ x | 0 | | 1 | |
|---|---|---|---|---|
| 0 | 1/2 | 0 | 1/2 | 0 |
|   | 0 | 1/2 | 0 | 1/2 |
| 1 | 1/2 | 0 | 0 | 1/2 |
|   | 0 | 1/2 | 1/2 | 0 |

Now we want to find a representation of this particular PR-box in terms of ordered vector spaces. Due to the fact that the dimension of the state space grows fast, we restrict to one half of this box. This is also called a gbit [53] (Table 3.1). Since the GPT formalism is strongly influenced by operationalism, one can think about the gbit as follows. A agent Alice holds a block box $\omega$ and she can choose between measurement $x \in \{0, 1\}$ and obtains an outcome $a \in \{0, 1\}$. As above the box is described by the conditional probability $p(a|x)$. Any state $\omega$ can be written as

$$\omega = (1, p(0|0), p(0|1)) \in \mathbb{R}^3 \qquad (3.17)$$

The first entry describes the normalization $p(0|0)+p(1|0) = p(0|1)+p(1|1) = 1$, i.e., for any input Alice or Bob receive with certainty an output. Since in principle all probabilities are allowed, this corresponds to the set

$$\mathbb{R}^3 \supset \Omega := \left\{ (1, t, s) \mid 0 \leq t, s \leq 1 \right\} \qquad (3.18)$$

Geometrically this set, or the state space $\Omega$, corresponds to a square. The extreme states are given by

$$\omega_1 = (1, 0, 0)^t \ , \ \omega_2 = (1, 1, 0)^t \ , \ \omega_3 = (1, 0, 1)^t \ , \ \omega_4 = (1, 1, 1)^t \qquad (3.19)$$

and every element in $\Omega$ can be written as a convex combination of $\omega_1, \ldots, \omega_4$. Every state on one side of the square is perfectly distinguishable from every state on the opposite site. Similar to the PR-box, we interpret Alice input $x$ as a choice of measurement, thus the two measurements are $e_{a=0}^{x=0}$, $e_{a=1}^{x=0}$ and $e_{a=0}^{x=1}$, $e_{a=1}^{x=1}$ such that $\sum_{a=0}^{1} e_a^x (\omega) = 1$ for all states in $\Omega$. By describing effects as vectors by using the standard inner product we obtain

$$e_{a=0}^{x=0} = (0, 1, 0)^t \ , \ e_{a=0}^{x=1} = (0, 0, 1)^t \ , \ e_{a=1}^{x=0} = (1, -1, 0)^t \ , \ e_{a=1}^{x=1} = (1, 0, -1)^t$$
$$(3.20)$$

For instance, one can easily check that $e_{a=0}^{x=0} (\omega) = \langle (0, 1, 0)^t , \omega \rangle = p\,(0|0)$. Since the unit effect is given by $u = (1, 0, 0)$ we obtain the other effects. In addition, there are four pure states labelled by $\omega_1, \ldots, \omega_4$. Every pure state $\omega_i$ is perfectly distinguishable from every other pure state $\omega_j$ for $j \neq i$, however no more than two of them are jointly distinguishable in a single measurement.

# Sections and Subsystems

**4**

## 4.1 Sections of the Effect Cone

As introduced in the previous chapter, a GPT $\mathfrak{A}$ is given by a tripel $\left(V, V^+, e_V\right)$, where $V^+$ contains all effects that can be implemented within the theory. In the framework of a general probabilistic description, a state is given by a linear functional, which is positive on the effect cone. Apparently the state space depends upon the chosen cone $V^+$ and is completely determined by $\mathfrak{A}$. A frequently used method to obtain new structures from old ones is the restriction to some subset. The aim of this chapter is to understand how new theories can emerge from old ones and to what extent the state space changes under this transformation.

**Definition 4.1** ([21]). Let $\mathfrak{A}_1 = \left(V, V^+, e_V\right)$ and $\mathfrak{A}_2 = \left(W, W^+, e_W\right)$ be two order unit spaces. We say that $\mathfrak{A}_1$ emerges as a section of $\mathfrak{A}_2$ if there exists a positive unital injection $\varphi : V \to W$ such that

$$\varphi\left(V^+\right) = W^+ \cap \operatorname{im}(\varphi) \tag{4.1}$$

It is important to note that this definition of a section only makes use of the space of observables and thus defines the transformation of state space implicit. Since $\varphi$ is given by an injective map, it follows that the dimension of $V$ is less or equal to the dimension of $W$. Given a physical system described by a GPT $\mathfrak{A}$, it could happen that all possible observables can be implemented. In this case, $\mathfrak{A}$ should be a section of itself. By choosing $\varphi = \operatorname{id}$ it is easy to check that (4.1) holds. Further if $\mathfrak{A}_1$ is a section of $\mathfrak{A}_2$ and $\mathfrak{A}_2$ is a section of another theory $\mathfrak{A}_3$ it is quite natural to assume that also $\mathfrak{A}_1$ emerges as an effective theory of $\mathfrak{A}_3$. In mathematical terms this means if $\varphi_1 : V_1 \to V_2$ and $\varphi_2 : V_2 \to V_3$ are sections, also the composition is a valid

J. Steinberg, *Extensions and Restrictions of Generalized Probabilistic Theories*, BestMasters, https://doi.org/10.1007/978-3-658-37581-2_4

section, i.e., $\varphi_2 \circ \varphi_1 =: \varphi_3 : V_1 \to V_3$ is a positive unital injection and fulfills (4.1). The unitality, injectivity and positivity of the composed map is obvious. To show the remaining, first notice that by assumption one can find left inverses to $\varphi_1$ and $\varphi_2$ denoted by $\varphi_1^{-1}, \varphi_2^{-1}$ respectively. Further $\varphi_1^{-1}$ and $\varphi_2^{-1}$ are positive and unital and therefore the composed is also a positive and unital map. This leads to the following diagram

$$\begin{array}{ccc} \mathfrak{A}_1 & \xrightarrow{\varphi_1} & \mathfrak{A}_2 \\ & \searrow{\scriptstyle\varphi_3} & \downarrow{\scriptstyle\varphi_2} \\ & & \mathfrak{A}_3 \end{array}$$

with $\varphi_3 = \varphi_2 \circ \varphi_1$. These simple observations have the consequence, that we can consider Archimedean order unit spaces together with sections as morphisms as subcategory of the category $\mathcal{O}_{Arch}$. We will denote this category by $\mathcal{O}_{Sec}$.

**Example 4.2** A first indication of the conceptual rigidity of sections is that a low dimensional quantum system should emerge as a section of a high dimensional quantum system. This can be seen as follows. Any quantum system can be characterized by the collections of observations that can be made on it through suitable measurement processes. The physical quantities that are thus accessed are the observables of the system, described within $\mathfrak{A}$ by $V^+$. Since a section is a positive unital injection one can interpret it as an improvement of the resolution of physical quantities. For instance, suppose the physical system under consideration is a spin-1 particle, i.e., an object with three levels. If the observer has only access to a set of measurements described by $\mathfrak{A}_1 = \left( H_2\left(\mathbb{C}\right), H^+\left(\mathbb{C}\right), \mathbb{I} \right)$ he can only distinguish between two different levels and thus observe a spin-$\frac{1}{2}$ particle. On the other hand if $\mathfrak{A}_2 = \left( H_3\left(\mathbb{C}\right), H^+\left(\mathbb{C}\right), \mathbb{I} \right)$ he can distinguish all three different levels. Assume we have a GPT $\mathfrak{A}_1$ describing a $m$-level quantum system and a second GPT $\mathfrak{A}_2$ describing a $n$-level quantum system with $s = n - m > 0$. Consider the map

$$\varphi : H_m\left(\mathbb{C}\right) \to H_n\left(\mathbb{C}\right) , \quad X \mapsto \frac{\operatorname{tr}\left(X\right)}{s} \mathbb{I}_s \oplus X \qquad (4.2)$$

where we have introduced the notation, that for given matrices $v \in M_m\left(V\right)$ and $w \in M_n\left(V\right)$ we declare

$$v \oplus w = \begin{pmatrix} v & 0 \\ 0 & w \end{pmatrix} \in M_{m+n}\left(V\right) \qquad (4.3)$$

where $V$ is a finite dimensional vector space. Since $X$ is assumed to be hermitian, also $\varphi(X)$ is hermitian in $H(n, \mathbb{C})$ such that $\varphi$ is well defined. It is important to note that the occurrence of the trace is not the consequence of some normalization procedure. Without the trace the map $\varphi$ would not be a homomorphism at all. On the other hand we have

$$\sigma\left(\varphi\left(X\right)\right) = \sigma\left(\frac{\operatorname{tr}\left(X\right)}{s}\mathbb{I}_s \oplus X\right) = \sigma\left(X\right) \cup \{\frac{\operatorname{tr}\left(X\right)}{s}\} \tag{4.4}$$

Therefore $\varphi$ is a positive unital injection. To verify that $\varphi$ is a valid section, we have to check whether $\varphi\left(\mathrm{H}_m^+\left(\mathbb{C}\right)\right) = \mathrm{H}_n^+\left(\mathbb{C}\right) \cap \varphi\left(\mathrm{H}_m\left(\mathbb{C}\right)\right)$. For this let $Y \in \mathrm{H}_n^+\left(\mathbb{C}\right) \cap \varphi\left(\mathrm{H}_m\left(\mathbb{C}\right)\right)$. Thus $Y$ must be positive and of the form

$$\begin{pmatrix} \lambda \mathbb{I}_s & 0 \\ 0 & Y_0 \end{pmatrix} \tag{4.5}$$

where we defined $\lambda = \operatorname{tr}\left(Y_0\right)/s$. Since this is a matrix of block form, we have $Y_0 \geq 0$ and $\lambda \geq 0$. Consequently we have $Y \in \varphi\left(\mathrm{H}_m^+\left(\mathbb{C}\right)\right)$. By construction we have $\varphi\left(\mathrm{H}_m^+\left(\mathbb{C}\right)\right) \subset \mathrm{H}_n^+\left(\mathbb{C}\right) \cap \varphi\left(\mathrm{H}_m\left(\mathbb{C}\right)\right)$ and thus $\varphi$ is a valid section. This is also reasonable from the physical perspective, since the measurement of $\mathbb{I}$ corresponds to the "do nothing measurement" on the ancilla system. As we already noticed, the aim of GPTs is to give a framework that allows us to describe a broad range of physical theories. In particular, it should unify quantum and classical theories. Therefore classical theories should emerge as sections of quantum theory. Classical theories are those ones, where the base of the cone is given by a simplex. Thus the extreme rays are given by a certain set of vectors $\mathbf{v}_i \in \mathbb{R}_+^n$ which can be embedded into $\mathrm{H}_n^+\left(\mathbb{C}\right)$ by virtue of $\iota\left(\mathbf{v}_i\right) = \operatorname{diag}\left(\left(\mathbf{v}_i\right)_{i=1}^n\right)$. It is important to note that the cone of any classical theory is given by the convex combination of its extreme rays. To see that $\iota$ is a section assume that the classical cone is generated by $\mathbf{v}_i$. But the set of diagonal matrices with real, positive entries are a proper subset of $\mathrm{H}_n^+$ and thus the claim follows.

At this point it is important to emphasize the analogies and differences between the GPT and the algebraic formalism of the description of a quantum system. In the algebraic framework the observables are regarded as a von-Neumann algebra $\mathcal{A}$ and a subsystem is described by a subalgebra $\mathcal{N}$. In the finite dimensional case we can simply think about $\mathcal{A}$ as the matrix algebra $\mathrm{M}_n\left(\mathbb{C}\right)$ and $\mathcal{N} \subset \mathcal{A}$ a matrix subalgebra. Further the set $\mathcal{K}$ of all positive semidefinite elements in $\mathcal{A}$ is a closed convex cone in $\mathcal{A}$ which is pointed, i.e., $\mathcal{K} \cap \left(-\mathcal{K}\right) = \{0\}$. In addition the set of hermitian matrices is a linear subspace of $\mathcal{A}$. Hence any observable algebra is in a natural way equipped with a cone $\mathcal{K}$ of positive elements. Consider again the spin-$\frac{1}{2}$ particle and the spin-1 particle. As we have already seen, we can regard the first one as a subsystem of the second in the GPT formulation. But it is not a subsystem in the sense of matrix algebras, i.e., there exists no unital algebra-homomorphism $\varphi$ such that

$$\varphi : M_2\left(\mathbb{C}\right) \to M_3\left(\mathbb{C}\right) \tag{4.6}$$

This indicates that if one describes quantum mechanics in terms of GPTs and sections, there are significantly more possibilities to introduce a subsystem as in the algebraic approach, i.e., the notion of sections is much more flexible. The relation between the description of a quantum system in terms of GPTs and the algebraic formulation becomes clear if one raises the level of abstraction, i.e., considers the category of sections $\mathcal{O}_{sec}$ and the category of quantum mechanics $\mathcal{O}_{QM}$. In $\mathcal{O}_{QM}$ the objects are given by von-Neumann algebras and the morphisms are completely positive maps. The correspondence can therefore be written as a functor $\mathcal{F} : \mathcal{O}_{QM} \to \mathcal{O}_{sec}$ acting by

$$\mathcal{F}\left(\mathcal{A}\right) = \left(\mathcal{A}_{sa}, \mathcal{K}, \mathbb{I}\right) \tag{4.7}$$

on objects, where $\mathcal{A}_{sa}$ denotes the linear subspace of self-adjoint elements of $\mathcal{A}$. Further its action on morphisms is given by

$$\mathcal{F}\left(\phi\right) : \mathcal{A}_{sa} \to \mathcal{B}_{sa} \tag{4.8}$$

such that

$$\mathcal{F}\left(\phi\right)\left(\mathcal{K}_{\mathcal{A}}\right) = \mathrm{im}\left(\mathcal{F}\left(\phi\right)\right) \cap \mathcal{K}_{\mathcal{B}} \tag{4.9}$$

**Example 4.3** Consider the order unit space given by $\left(\mathbb{R}^{n+1}, C_{n+1^+}\left(c\right), \left(1, \mathbf{0}\right)\right)$ where $C_{n+1}^+\left(c\right)$ denotes the $(n+1)$-dimensional Euclidean cone with length-diameter ratio $\frac{c}{2}$ that is

$$C_{n+1}^+\left(c\right) = \{(t, \mathbf{x}) \in \mathbb{R}_+ \times \mathbb{R}^n \mid t \geq c||\mathbf{x}||_2\} \tag{4.10}$$

To show that this space is a section of quantum mechanics, we have to make use of the structure of Clifford algebras. For $n \in \mathbb{N}$ we call the free real associative algebra generated by $\Gamma_1, ..., \Gamma_{2n}$, subject to the anti-commutation relations

$$\{\Gamma_j, \Gamma_k\} = \Gamma_j \Gamma_k + \Gamma_k \Gamma_j = 2\delta_{ij}\mathbb{I} \tag{4.11}$$

Clifford algebra. Up to unitary equivalence, the Clifford algebra has a unique representation by hermitian matrices on $n$ qubits. If one fixes a basis, this representation can be obtained by the Jordan-Wigner transformation [58]

$$\Gamma_{2j-1} = \sigma_z^{\otimes(j-1)} \otimes \sigma_x \otimes \mathbb{I}^{\otimes(n-j)}, \; \Gamma_{2j} = \sigma_z^{\otimes(j-1)} \otimes \sigma_y \otimes \mathbb{I}^{\otimes(n-j)} \tag{4.12}$$

for $j = 1, ..., n$ and $\otimes$ the Kronecker product of matrices. Following [21, 59] we define the map

$$\varphi : \mathbb{R}^{n+1} \to \mathrm{H}\left(2^m, \mathbb{C}\right) \;,\; (t, \mathbf{x}) \mapsto t\mathbb{I} + c\sum_j x_j \Gamma_j \tag{4.13}$$

It is clear that $\varphi$ is injective, linear and unital. It remains to show that $\varphi$ is positive and the existence of a positive unital left-inverse $\varphi^{-1}$. Since the $\Gamma$ matrices can be seen as a generating set for a Clifford algebra, we can make use of the anti-commutation relations (4.11) and obtain

$$2t\varphi(t, \mathbf{x}) = t^2\mathbb{I} + 2tc\sum x_j \Gamma_j + \frac{c^2}{2}\sum_{i \leq j} x_i x_j \left(\Gamma_i \Gamma_j + \Gamma_j \Gamma_i\right) \tag{4.14}$$

$$= t^2\mathbb{I} + 2tc\sum_j x_j \Gamma_j + c^2 ||\mathbf{x}||_2^2 \mathbb{I}$$

This can be rewritten as $2t\varphi(t, \mathbf{x}) = \varphi(t, \mathbf{x})^2 + \left(t^2 - c^2||\mathbf{x}||_2^2\right)\mathbb{I}$. Since $\varphi$ is an injection, choose as a left inverse

$$\varphi^{-1} : \mathrm{H}\left(2^m, \mathbb{C}\right) \to \mathbb{R}^{n+1} \;,\; A \mapsto \frac{1}{2^m}\left(\mathrm{tr}\,(A), \mathrm{tr}\,(A\Gamma_1)/c, ..., \mathrm{tr}\,(A\Gamma_{2m})/c\right) \tag{4.15}$$

Indeed $\varphi^{-1}$ is also positive and unital. To see this, define $x_i \equiv \mathrm{tr}\,(A\Gamma_i)$. But

$$\mathrm{tr}\,(A) - c||\mathbf{x}/c||_2 = \mathrm{tr}\,(A\varphi(1, -(\mathbf{x}/||\mathbf{x}||_2)/c)) \geq 0 \tag{4.16}$$

since both matrices in the trace are already positive.

**Theorem 4.4** Let $\mathfrak{A} = (V, V^+, e)$ be an order unit space which is complete in its order unit norm. The canonical representation $(\rho, S)$ and the smallest separating functional representation are sections of the GPT $\mathfrak{A}$.

*Proof.* By Theorem 2.31 the canonical representation $(\rho, S)$ of an order unit space $(A, e)$ over its state space is a functional representation which is the largest one separating points. Clearly $\rho(e) = 1_S$ such that $\rho$ is unital. Further by definition of $\rho$, its range $\rho(A)$ is given by all restrictions to $S$ of evaluation functionals $f \mapsto f(a)$ for some $a \in A$ and $f \in A^*$. These are $w^*$-continuous linear functionals on $A^*$. In addition any order unit space $(A, e)$ admits a functional representation and by virtue of Lemma 2.30 $\rho$ is injective and in both ways order preserving. Thus, $\rho$ is a section.                                                                                                       $\square$

Apart from various applications in mathematics the Kadison theorem also gains a physical interpretation by virtue of Theorem 4.4. In particular Theorem 2.31 assures that any theory can be simulated by a classical theory if the dimension of the state space is sufficiently high.

**Theorem 4.5** Let $(V, V^+, e_V)$ and $(W, W^+, e_W)$ be Archimedean order unit spaces. The map

$$\varphi : V \to V \otimes_{\max} W \quad , \quad v \mapsto v \otimes e_W \tag{4.17}$$

is section.

*Proof.* Obvious the map $\varphi$ is unital since $e_V \otimes e_W$ is an Archimedean order unit for $V \otimes_{\max} W$. If $v \in V^+$ then $\varphi(v) = v \otimes e_W$ and since $e_W \in W^+$ we have $\varphi(V^+) \subset V^+ \otimes W^+ \subset D$. Hence $\varphi$ is unital, positive. The injectivity follows from the definition of the tensor product. It remains to construct a positive left inverse of $\varphi$. Consider the map

$$\Psi_\omega : V \otimes_{\max} W \to V \quad , \quad \Psi_\omega(u) := (\mathrm{id}_V \otimes \omega)(u) \tag{4.18}$$

Since we can expand any element $u \in V \otimes_{\max} W$ in terms of elementary tensors we have $\Psi_\omega(u) = \Psi_\omega\left(\sum_{i=1}^n a_i \otimes b_i\right) = \sum_{i=1}^n \Psi_\omega(a_i \otimes b_i) = \sum_{i=1}^n \omega(b_i) v_i$. Since $\Psi_\omega$ is linear it is independent of decomposition and because $\omega$ is a state, it is also unital. To show positivity choose $u \in D$ and remember that $D$ is defined as

$$D := \left\{ z \in V \otimes W : \forall \epsilon > 0,\, z + \epsilon e_V \otimes e_W \in V^+ \otimes W^+ \right\} \qquad (4.19)$$

Then

$$\Psi_\omega (u) = \Psi_\omega (z + \epsilon e_V \otimes e_W) = \Psi_\omega (z) + \epsilon e_V \qquad (4.20)$$

By construction of $D$ we have $z + \epsilon e_V \otimes e_W \in V^+ \otimes W^+$ for all $\epsilon > 0$. Thus we can rewrite $u$ as $z + \epsilon e_V \otimes e_W = \sum_{i=1}^{m} v_i \otimes w_i$ where $v_i \in V^+$ and $w_i \in W^+$. This implies

$$\Psi_\omega (u) = \Psi_\omega \left( \sum_{i=1}^{m} v_i \otimes w_i \right) = \sum \omega (w_i)\, v_i \in V^+ \qquad (4.21)$$

since $\omega \in S(W)$. But $\left( V, V^+, e_V \right)$ was Archimedean and so $\Psi_\omega (z) + \epsilon e_V \in V^+$ for all $\epsilon > 0$ implies $\Psi_\omega (z) \in V^+$. Since $V^+$ is a cone, also $\Psi_\omega (u) = \Psi_\omega (z) + \epsilon e_V \in V^+$, what completes the proof. But Lemma 4.6 tells us that any positive unital linear injection with a positive inverse is a section. $\qquad \square$

## 4.2    Sections of the State Space

Assume $\left( V, V^+, e_V \right)$ and $\left( W, W^+, e_W \right)$ are Archimedean order unit spaces and $\varphi : V \to W$ is a section of those. Further we denote by $S(V)$ and $S(W)$ the state space of $V$ and $W$ respectively. Since the state space is a convex subset of the dual space, the map $\varphi$ induces the map $\varphi^* : W^* \to V^*$ with $W^* \ni w \mapsto \varphi^* (w) = w \circ \varphi$. This may allow two different constructions of the state space of the section $\left( V, V^+, e_V \right)$. In particular, the direct definition of the state space by the set of all positive, unital linear functionals on $V$ or by using the induced map, i.e.,

$$\tilde{S}(V) := \left\{ \sigma \circ \varphi \mid \sigma \in S(W) \right\} = \varphi^* (S(W)) \qquad (4.22)$$

At this point the question arises whether or not both definitions are equivalent, that is $\varphi^* (S(W)) = S(V)$.

**Lemma 4.6** Let $\mathfrak{A}_1$ and $\mathfrak{A}_2$ be Archimedean order unit spaces and $\varphi : V \to W$ a section, i.e., a linear unital map fulfilling $\varphi \left( V^+ \right) = W^+ \cap \mathrm{im}\, (\varphi)$. Then there exists a map $\Psi : W \to V$ with $\Psi \circ \varphi = \mathrm{id}_V$ which is also unital and positive, i.e., $\Psi (e_W) = e_V$ and $\Psi \left( W^+ \right) \subset V^+$.

*Proof.* The proof can be divided into two parts. First assume that $V \cong W$. Since $V^+$ is a full cone in $V$ and $\varphi$ is an injection we have im $(\varphi) = W$. Thus $\varphi(V^+) = W^+ \cap W = W^+$ and clearly $\Psi(W^+) = V^+$. In particular, the map $\Psi$ is positive. If dim $(V) = m < n = $ dim $(W)$, then $\varphi(V)$ is a proper linear subspace of $W$. When $\mathcal{B}_1 = \{\alpha_1, ..., \alpha_m\}$ is a basis for $\varphi(V)$ we can complete it to a basis of $W$ by adding $\{\alpha_{m+1}, ..., \alpha_n\}$, i.e., $\mathcal{B}_2 = \{\alpha_1, ..., \alpha_n\}$ such that $\langle \mathcal{B}_2 \rangle = W$. The map $\Psi$ is by assumption on $\varphi(V)$ defined as a left inverse of $\varphi$, hence it is fixed on the elements of $\mathcal{B}_1$ and extents by linearity to the whole linear subspace. On $\mathcal{B}_2 \setminus \mathcal{B}_1 = \{\alpha_{m+1}, ..., \alpha_n\}$ we fix $\Psi$ to zero. This can be done since the elements of $\mathcal{B}$ are pairwise linearly independent. It remains to check whether or not this construction leads to a positive left inverse. Indeed $\Psi(W^+) = \Psi(W^+ \cap \varphi(V))$. At this point it could in principle happen that $W^+ \cap \varphi(V)$ contains elements of $\varphi(-V)$. But we have $\varphi(V^+) = W^+ \cap \varphi(V)$ and therefore $\Psi(W^+) = \Psi(\varphi(V^+)) = V^+$. $\qquad\square$

**Theorem 4.7** Let $\mathfrak{A}_1$ and $\mathfrak{A}_2$ be Archimedean order unit spaces and $\varphi : V \to W$ a section, i.e., a linear unital map fulfilling $\varphi(V^+) = W^+ \cap$ im $(\varphi)$. Then $\tilde{S}(V) = S(V)$ holds.

*Proof.* Since $\varphi$ is unital and order preserving, $\omega \in S(W)$ implies $\omega \circ \varphi \in S(V)$. Therefore we obtain the inclusion $\tilde{S}(V) \subset S(V)$. Further Lemma 4.6 guarantees the existence of a left inverse $\Psi$ to $\varphi$ such that $\Psi$ is also a positive map. Let $\sigma \in S(V)$. Clearly $\sigma = \sigma \circ$ id $= \sigma \circ \Psi \circ \varphi$. Since $\Psi$ is unital and the composition of sections is associative, $\omega := \sigma \circ \Psi$ defines a state on $W$. Hence we can write $\sigma = \omega \circ \varphi$. But $\sigma \in \tilde{S}(V)$, what implies $S(V) = \tilde{S}(V)$. $\qquad\square$

It is important to note that the proof only used the fact that $\Psi$ is positive. This raises the questions whether the condition $\varphi(V^+) = W^+ \cap$ im $(\varphi)$ is too strong. Equivalently formulated, do there exist maps $\varphi$ and $\Psi$ with $\Psi \circ \varphi = $ id such that both are positive but the condition (4.1) does not hold. Let $\varphi$ and $\Psi$ be positive unital maps such that $\Psi$ is a left inverse of $\varphi$. In particular, $\Psi(W^+) \subset V^+$ and $\varphi(V^+) \subset W^+$. It follows $\Psi(\varphi(V^+)) = V^+ \subset \Psi(W^+) \subset V^+$ thus $\Psi(W^+) = V^+$. Assume $\varphi(V^+) \subsetneq W^+ \cap$ im $(\varphi)$. This implies the existence of $v \in V \setminus V^+$ such that $\varphi(v) \in W^+$. But $\Psi$ is a positive left inverse, hence $\Psi(W^+) = V^+$ and therefore $\Psi(\varphi(v)) \in V^+$ what is a contradiction.

Since a GPT $(V, V^+, e)$ and its state space S are dual objects, one is in principle free to choose whether the order unit space or the states $S \subset U$ with some embedding vector space $U$ is the fundamental object of the theory. If $S$ is the fundamental object, then we can define $V$ to be the space of affine functions on $U$, let $V^+ =$

$\{\xi \in V | \xi \, (\mathrm{S}) \subset \mathbb{R}^+\}$ and choose $e$ with $e \, (\mathrm{S}) = \{1\}$. To guarantee that both objects can serve as a fundamental one, there should exist an isomorphism between $V$ and $V^{**}$, where $V^{**}$ denotes the algebraic double dual space. If $V$ is finite dimensional it follows that $V \cong V^{**}$ and the algebraic double dual $V^{**}$ coincides with the topological dual $V''$. In particular, this canonical isomorphism $\Phi$ is simply given by

$$\Phi : V \to V^{**} \, , \quad \Phi \, (v) = \iota_v \text{ with } \iota_v : V^* \to \mathbb{R} \, , \quad \iota_v \, (\varphi) = \varphi \, (v) \qquad (4.23)$$

However, in the case of infinite dimensional vector spaces, $V^{**}$ and $V''$ do not coincide since there can exist discontinuous linear maps. In the following we restrict to $V''$. Also here there exists a natural homomorphism that becomes a bijection if $V$ is reflexive. Let $\mathrm{S}_T \subset T$ be a state space embedded into a vector space $T$ and let $\mathrm{S}_U \subset U$ be a state space embedded into a vector space $U$. Clearly it is our goal to define sections of the state space in such a way, that its dual can serve as a section with respect to Definition 4.1. To do so we can apply Theorem 2.28, telling us that the dual of an order unit space is a base norm space and vice versa. Thus we can restrict to the case of base norm spaces. Consider two base norm spaces $(V^*, S_V)$ and $(W^*, S_W)$ and the mapping

$$\left(W^*, S_W\right) \xrightarrow{\varphi^*} \left(V^*, S_V\right) \rightsquigarrow \left(V^*, S_V\right)^* \xrightarrow{\varphi^{**}} \left(W^*, S_W\right)^* \qquad (4.24)$$

where the action of the induced double dual map $\varphi^{**}$ on $v \in V^{**}$ is given by $\varphi^{**} \, (v) = v \circ \varphi$. The resulting functional $\varphi^{**} \, (v)$ is called the pullback of $v$ along $\varphi^*$. If $\varphi : V \to W$ is a surjective map, then its dual $\varphi^* : W^* \to V^*$ is clearly injective since $\omega \circ \varphi = 0$ implies $\omega = 0$ for $\omega \in W^*$. Hence we assume that the map $\varphi^*$ is surjective such that $\varphi^{**}$ is injective. By identifying the double dual spaces with the spaces itself, we also identify $\varphi^{**}$ with $\varphi$ and thus $\varphi$ is injective as claimed in Definition 4.1. Apart from requirement that $\varphi^*$ is surjective we further impose that $\varphi^* \, (S_W) \subset S_V$, i.e., $\varphi^*$ maps states to states. Assume that $\left(V, V^+, e_V\right)$ and $\left(W, W^+, e_W\right)$ are Archimedean order unit spaces. The dual spaces are then given by base norm spaces of the form $(V^*, S_V)$ and $(W^*, S_W)$.

**Theorem 4.8** Let $\mathfrak{A}_1 = \left(V, V^+, e_V\right)$ and $\mathfrak{A}_2 = \left(W, W^+, e_W\right)$ be Archimedean order unit spaces and denote by $(V^*, S_V)$ and $(W^*, S_W)$ their dual spaces. Further let

$$\varphi^* : W^* \to V^* \, , \quad \omega \mapsto \varphi^* \, (\omega) \text{ where } \varphi^* \, (\omega) \, (v) = (\omega \circ \varphi) \, (v) \qquad (4.25)$$

with $v \in V$, $\varphi^*(S_W) = S_V$ and $\varphi^*$ an epimorphism. Then the map $\varphi : V \to W$ is a section.

*Proof.* Let $v \in V^+$ and $\omega \in S_W$. Clearly

$$\omega(v) = \omega \circ \varphi(v) = (\omega \circ \varphi)(v) = \varphi^*(\omega)(v) \geq 0 \qquad (4.26)$$

Further we obtain by virtue of $(\omega \circ \varphi)(e_V) = \varphi^*(\omega)(e_V) = 1$ that $\varphi$ is also unital. It remains to check whether $\varphi(V^+) = W^+ \cap \mathrm{im}(\varphi)$ holds. Suppose there is $v \notin V^+$ such that $\varphi(v) \in W^+$. Consequently one has

$$\varphi^*(\omega)(v) = (\omega \circ \varphi)(v) \geq 0 \ \forall \omega \in S_W \qquad (4.27)$$

But $\omega \circ \varphi$ is a state on $V$ and by assumption we have that $\{\omega \circ \varphi : \omega \in S_W\} = S_V$. But by Theorem 2.23 we obtain that $v \in V^+$. Contradiction. $\qquad \square$

This allows us to reformulate Definition 4.1 for the case that the states are taken as fundamental objects.

**Definition 4.9** Let $(V, K_1)$ and $(W, K_2)$ be base norm spaces. We call a map $\varphi : (W, K_2) \to (V, K_1)$ a section, if $\varphi$ is a epimorphism and $\varphi(K_2) = K_1$.

## 4.3   Spekkens' Toy Model

The Spekkens' model [60] is a toy theory with the aim of showing that many characteristic features of quantum mechanics could result from a restriction on our knowledge of the state of an essentially classical system. The toy theory describes a simple type of system which mimics many of the features of a quantum qubit. While quantum mechanics represents states of systems by vectors in a Hilbert space and processes are represented by linear maps, the toy theory represents states by subsets and processes by relations. A crucial concept is the dichotomy between states of reality and states of knowledge. States of reality are called ontic whereas states of knowledge are called epistemic. Note that an epistemic state is simply a subset of the ontic state space. The toy theory now treats all quantum states, mixed and pure, as states of incomplete knowledge. Hence the toy model represents a hidden-variables model. Spekkens' theory relies on the knowledge balance principle stating that if one has maximal knowledge, then for every system, at every time, the amount of knowledge one possesses about the ontic state of the system at that time must equal the amount of knowledge one lacks. This implies that only certain epistemic states are allowed. But to give this principle a meaning, one has to make the notion of knowledge more concrete. A canonical set of yes/no questions is a set of yes/no questions that is sufficiently to fully specify the ontic state and that has a minimal number of elements. Note that this set is not uniquely determined. One defines the measure of knowledge as the maximum number of questions for which the answer is known, in a variation over all canonical set of questions. In analogy the measure of ignorance is defined as the difference between this number and the total number of questions in the canonical set. Since the canonical set is the minimal sufficient set of questions to determine the ontic state, it follows that for two questions there are four possible ontic states. Hence the simplest possible system within the toy theory has four ontic states and is called an elementary system. Given a composite system of $n$ elementary systems, its ontic state space is simply the cartesian product of the ontic state spaces of the composite systems. Thus such system has $2^{2n}$ ontic states and $2n$ questions on the canonical set. For an elementary system we have two questions in the canonical set and four ontic states. According to the knowledge balance principle, the answer of only one of these can be known. The epistimic states for which the balance occurs are those which identify the ontic state of the system to be one of two possibilities. If one denotes the ontic state space by $\mathcal{A}_4 = \{1, 2, 3, 4\}$ and the disjunction by the symbol $\vee$, the six epistemic states of maximal knowledge can be represented as in Figure 4.1. To define an analogy of convex combination of states, one needs the concept of an ontic base and disjointness. The ontic base of an epistemic state is the set of ontic states which are consistent with it. If the intersection of the ontic bases of a pair of epistemic states is empty, then those states are said to be disjoint. The combination of epistemic states is only declared for disjoint states that whereby in addition the union of their ontic basis must the ontic base of a valid epistemic state. Note that the definition of the convex combination of a set of epistemic states differs from the convex combination of a set of quantum states since there is no analogous to a convex sum with unequal weights. An epistemic state is called mixed if one can obtain this state as a convex combination of distinct epistemic states. Otherwise the state is called pure. Further we have to equip the toy theory with sorts of transformations on the ontic states that are allowed by the knowledge balance principle. It turns out that the knowledge balance principle forbid many-to-one maps and therefore one has to consider one-to-one or one-to-many maps. Since the former correspond to reversible maps, one focus on them. Obvious, these are simply the set of permutations of the four ontic states.

**Figure 4.1** Graphical description of the epistemic states of maximal knowledge. The four cells representing the four ontic states and the filled cells denote the set in which the actual ontic state of the system is known to lie [60, 61]

For the description of measurements, one shall consider measurements that are reproducible, i.e., if repeated upon the same system, they yield the same outcome. For this to be possible, the epistemic state after the measurement must rule out all of the ontic states that are not consistent with the outcome. The knowledge balance principle imposes restrictions on the sort of reproducible measurements that can be implemented. In analogy to the allowed transformations of states, one can rule out a certain kind of measurements, namely those ones which identifies whether or not the ontic state is in a singleton set.

**Spekkens' theory as a QM section**
Also the Spekkens' model can be formulated within the framework of ordered vector spaces [62]. The positive cone is embedded into $\mathbb{R}^4$ and is generated by the six extremal elements $\partial^+ V^+ = \{a_{\pm 1}, a_{\pm 2}, a_{\pm 3}\}$ given by $a_i = \left(\frac{1}{2}, \mathbf{a}_{(i)}\right)$ with

$$\mathbf{a}_{(\pm 1)} = \left(\pm\frac{1}{2}, 0, 0\right), \mathbf{a}_{(\pm 2)} = \left(0, \pm\frac{1}{2}, 0\right), \mathbf{a}_{(\pm 3)} = \left(0, 0, \pm\frac{1}{2}\right) \qquad (4.28)$$

such that the order unit is $e = a_{+k} + a_{-k} = (1, 0)$. It is clear that the positive cone of the theory is given by the conic combination of the six extremal elements. Further we use the abbreviation **Spek** when we refer to this GPT. Remember that for a $n$-level quantum system, the corresponding GPT is given by $\left(H_n, H^+, \mathbb{I}\right)$. We are now interested in the following optimization task

$$\min_{n \in \mathbb{N}} \left(\exists \varphi \text{ section} \mid \varphi : \mathbf{Spek} \to \mathrm{QM}_n\right) \qquad (4.29)$$

Since $\varphi$ is an injective map, the case $n = 1$ is trivially excluded. However, since the dimensions of $M_2$ ($\mathbb{C}$) and $\mathbb{R}^4$ agree, we choose $n = 2$ and hence implement a qubit system.

**Lemma 4.10** The Spekkens' model cannot be obtained as a section of a qubit system.

*Proof.* Suppose there exists a section $\phi$ : **Spek** $\rightarrow$ QM$_2$. Since the cone $V^+$ is generated by its extremal elements we have

$$\phi\left(V^+\right) = \phi\left(\text{cone}\,(a_{\pm i})_{i=1}^3\right) = \text{cone}\left(\phi\,(a_{\pm i})_{i=1}^3\right) \tag{4.30}$$

Since we assumed that $\phi$ is injective, $\phi\left(V^+\right)$ is s a cone with at most six extremal elements. Equivalently formulated any base of the cone $\phi\left(V^+\right)$ is a polytope in the ambient space $\mathbb{R}^3$ with at most six vertices. On the other hand, the cone H$_2^+$ is generated by all rank-one projectors. But these projectors are exactly the surface of the Bloch ball. Hence this cone is generated by infinitely many extremal points and one can take $\partial B$ as a base of this cone. Since im $(\phi)$ is a linear subspace of M$_2$ ($\mathbb{C}$) constrained by $\mathbb{I} \in$ im $(\phi)$, the intersection H$_2^+ \cap$ im $(\phi)$ is cone generated by infinitely many extremal points. Thus, the both sets cannot coincide. $\qquad\square$

Clearly the proof of Lemma 4.10 makes use of the geometry of the state space. In particular, we used that the qubit state space is given by the Bloch ball, or more precise, that it does not contain any straight segment. This in principle encodes, why the optimization task (4.29) is a hard problem. Apart from the algebraic characterization of the state space of a quantum system, the geometric properties are widely unknown for $n \geq 3$. Referring to our problem, it is unknown whether the qudit state space for $d \geq 3$ contains a four dimensional facet. Fortunately, the Kadison representation Theorem 2.31 gives us an upper bound on $n$. Before we can obtain this upper bound, we have to characterize the states in the Spekkens' model.

**Lemma 4.11** The state space of **Spek** is given by convex hull of $\{\omega_1, ..., \omega_8\}$ where $\omega_i = (1, \boldsymbol{\omega}_i)$ with

$$\boldsymbol{\omega}_i = (a, b, c) \quad \text{and} \quad (a, b, c) \in \{-1, 1\}^3 \tag{4.31}$$

*Proof.* The general recipe to calculate the state space from a given observable cone $\mathcal{K} \subset \mathbb{R}^n$ is as follows. First one has to compute the dual cone

$$\mathcal{K}^* = \left\{y \in \mathbb{R}^n | \langle x, y \rangle \geq 0 \,\forall x \in \mathcal{K}\right\} \tag{4.32}$$

Since the state space is a base norm space, we have to project the dual cone onto the hyperplane defined by $\langle e, y \rangle = 1$ for all $y \in \mathcal{K}^*$. For the second task one has in general to use Fourier-Motzkin elimination [63]. Thus, computing the state space for a given cone is a computationally inefficient problem. However, computing the state space for **Spek** is due to the simplicity of the model easy to tackle. First, let $y = (y_1, y_2, y_3, y_4)^t \in \mathbb{R}^4$. Since the coordinates of the vectors are given, we can replace $\langle x, y \rangle = x^t y$. From the normalization condition we obtain directly $y_1 = 1$. Because $y_1$ is already fixed, any pair of extreme rays of the cone $\mathbf{a}_{(\pm i)}$ gives a system of independent inequalities which can be solved separately.                                   $\square$

**Theorem 4.12**  The Spekkens' model **Spek** can be obtained as a section of a 8 level quantum system.

*Proof.*  By Lemma 4.11 the state space of **Spek** is the convex hull the six extremal elements $(\omega_i)_{i=1}^8$. By virtue of Theorem 2.31 there exists a functional representation of **Spek** over its state space. In particular, the smallest representation is given by

$$\varphi : \mathbf{Spek} \to C_{\mathbb{R}} (\omega_1, ..., \omega_8) \tag{4.33}$$

Since we have to regard $\partial_e S = \{\omega_1, ..., \omega_8\}$ as a locally compact Hausdorff space, all finite sets are closed, thus all subsets are closed as countable union of closed sets. Therefore the only topology on $\partial_e S$ is the discrete topology. Hence we can identify $C_{\mathbb{R}} (\omega_1, ..., \omega_8)$ with $\mathbb{R}^8$ in the canonical way, i.e., both spaces are also topologically equivalent. In the following we call this canonical identification $\beta$. In Example 4.2 we have already proven that any classical system is by virtue of the map $\iota$ a section of a quantum system. Since the composition of sections is again a valid section the claim follows, i.e., the map $\iota \circ \beta : \mathbf{Spek} \to QM_8$ is a section.                              $\square$

# Two-Sections of Quantum Mechanics 5

As we have already seen in the previous chapter it is a difficult task to decide whether Spekkens toy model is a section of quantum mechanics or not. This difficulties occur due to some lack of knowledge about the geometry of the state space in quantum mechanics. In this chapter we first present some features of the state space of two and three level systems. Then we give sufficient conditions for the classification of the geometry of two dimensional subspaces that stem from an $n$-level quantum system.

## 5.1 Classification of Two Dimensional Sections

As we have already seen, thanks to the inner product structure of the space of self adjoint operators, we can identify states with density matrices, i.e., matrices of the form

$$\Omega_n = \left\{ \rho \in H_n\left(\mathbb{C}\right) \mid \rho \geq 0,\ \mathrm{tr}\left(\rho\right) = 1 \right\} \tag{5.1}$$

The dimension of this convex set is defined as the dimension of the vector space generated by this elements. Since the dimension of H $(n, \mathbb{C})$ is $n^2$ and the trace condition defines an affine subspace, one obtains $n^2 - 1$ for the dimension of $\Omega_n$. If the dimension is finite it is clear that $\Omega_n$ is also compact. Before we present classification statements for the geometry of subspaces of state spaces, let us first review what is known for qubits and qutrits. In fact, it turns out that apart from this two cases only a very little is known.

© The Author(s), under exclusive license to Springer Fachmedien Wiesbaden GmbH, part of Springer Nature 2022
J. Steinberg, *Extensions and Restrictions of Generalized Probabilistic Theories*, BestMasters, https://doi.org/10.1007/978-3-658-37581-2_5

## Qubit

For qubits, the states are precisely given by the Bloch ball and the pure states are located at the sphere. Thus the topological extreme points coincide with the convex ones. The cone of observables for the qubit are all $2 \times 2$ hermitian matrices and so a linear space of dimension 4. The unit matrix $\sigma_0 \equiv \mathbb{I}$ along with the Pauli matrices $\sigma_1, \sigma_2, \sigma_3$ form a complete set of hermitian orthogonal matrices. Any $2 \times 2$ hermitian matrix can be uniquely expressed as a linear real combination, i.e., $\rho(\mathbf{x}) = \frac{1}{2}(\sigma_0 + \mathbf{x}\sigma)$ for $\mathbf{x} \in \mathbb{R}^3$. Another characteristic that discriminate the case $n = 2$ from higher dimensional state spaces is how a unitary evolution acts on it. Any $U \in SU(2)$ defines a unitary evolution by $\rho \mapsto \tilde{\rho} := U\rho U^*$ which acts on $\Omega_2 \subset \mathbb{R}^3$ through $SO(3)$ rotations arising from the adjoint representation of $SU(2)$.

## Qutrit

Now we pass from the qubit to the qutrit case. The qutrit is a 3-level quantum system and hence the pure states can be described by a three dimensional Hilbert space. In the computational basis this three orthogonal pure states are denoted by $|0\rangle, |1\rangle, |2\rangle$. As usual there is a one-to-one correspondence between the states of the quantum system and positive semidefinite unit-trace operators. We denote the collection of all states by $\Omega_3$. It turns out that the structure of the state space of qutrits is much more richer than for qubits, where it is simply the Bloch ball. Since we want to know whether or not Spekkens' toy model emerge as a section of the qutrit, a good understanding of the geometry of $\Omega_3$ is important. In the qubit case we had to deal with two obstructions. First that the state space, i.e., the Bloch sphere is uniformly convex with respect to the norm induced by the inner product $(A, B) := \text{Tr}(AB)$ and the state space of the Spekkens' model is not. Second we cannot find a linear subspace such that the intersection contains as much extreme rays as Spekkens' to model. The hermitian trace class matrices can be described by

$$\rho(\mathbf{x}) = \frac{1}{3}\left(\mathbb{I} + \sqrt{3}\mathbf{x}\lambda\right) \quad \text{where} \quad \mathbf{x} \in \mathbb{R}^8 \tag{5.2}$$

The $\rho(\mathbf{x})$ are hermitian since the $\lambda$-matrices are the generators of $SU(3)$ and the coefficients are taken from $\mathbb{R}$. The prefactor guarantees the unit trace property. It remains to choose $\mathbf{x}$ in such a way that $\rho$ is positive semidefinite. If $\Omega_3$ contains all $\mathbf{x} \in \mathbb{R}^8$ such that $\mathbb{I} + \sqrt{3}n\lambda \geq 0$ it provides by definition the state space. Following [64] the state space can be parameterized via

$$\Omega_3 = \left\{\mathbf{x} \in \mathbb{R}^8 \mid 3\mathbf{x} \cdot \mathbf{x} - 2(\mathbf{x} * \mathbf{x}) \cdot \mathbf{x} \leq 1, \ \mathbf{x} \cdot \mathbf{x} \leq 1\right\} \tag{5.3}$$

where $* : \mathbb{R}^8 \to \mathbb{R}^8$. In the same way one can characterize the topological and the convex extreme points

$$\partial \Omega_3 = \left\{ \mathbf{x} \in \mathbb{R}^8 \mid 3\mathbf{x} \cdot \mathbf{x} - 2 (\mathbf{x} * \mathbf{x}) \cdot \mathbf{x} = 1 \; , \; \mathbf{x} \cdot \mathbf{x} \le 1 \right\} \tag{5.4}$$

$$\Omega_3^{\text{ext}} = \left\{ \mathbf{x} \in \mathbb{R}^8 \mid \mathbf{x} \cdot \mathbf{x} = 1 \; , \; \mathbf{x} * \mathbf{x} = \mathbf{x} \right\} \tag{5.5}$$

By comparing (5.4) and (5.5) it follows that $\Omega_3^{\text{ext}} \subsetneq \partial \Omega_3$. This is a clear difference to the qubit case, where every convex extreme point is also extreme in the topological sense. Geometrically this means that the boundary of the state space $\Omega_3$ contains straight segments.

**Lemma 5.1** Let $A, B \in H_n (\mathbb{C})$ traceless and orthonormal and consider the map

$$f : \mathbb{R}^2 \to H_n (\mathbb{C}) \quad , \quad f (x, y) = xA + yB + \mathbb{I} \tag{5.6}$$

Define the spectrum of the map $f$ via $\sigma (f) = \sigma (x, y) = \sigma (xA + yB + \mathbb{I})$. The eigenvalues of $f$ are linear functions in $x$, $y$ if and only if $A$ and $B$ commute.

*Proof.* First observe that $\mathbb{I}$ commutes with every matrix, in particular with $g (x, y) = xA + yB$. Hence $\sigma (f) = \{\lambda_i + 1 \mid \lambda_i \in \sigma (g)\}$ and it is sufficient to consider the map $g$. If $A$ and $B$ commute, then one can find $U \in \mathcal{U} (n, \mathbb{C})$ such that both are diagonal. Thus we obtain the equivalent equation $xD_A + yD_B$, where $D_A$ and $D_B$ are the corresponding diagonal matrices with real entries. But then $g$ is in diagonal form and the eigenvalues are given by $\sigma (g) = \{a_1 x + b_1 y, ..., a_n x + b_n y\}$ with $a_i \in \sigma (A)$ and $b_i \in \sigma (B)$. To show the converse note that also $xA + yB$ is hermitian and one can find $V \in \mathcal{U} (n, \mathbb{C})$ such that $g (x, y) = \text{diag} (d_1 (x, y), ..., d_n (x, y))$ and the $d_i$ are by assumption linear functions in $x$, $y$. Thus we can decompose the $d_i$ as $d_i (x, y) = d_i^{(1)} x + d_i^{(2)} y$ with $d_i^{(1)}, d_i^{(2)} \in \mathbb{R}$. This implies

$$g (x, y) = x \cdot \text{diag} \left( d_1^{(1)}, ..., d_n^{(1)} \right) + y \cdot \text{diag} \left( d_1^{(2)}, ..., d_n^{(2)} \right) =: xA + yB \tag{5.7}$$

But $A, B$ are both diagonal matrices and so they commute. $\qquad \square$

Note that the requirement of being orthonormal can in principle be replaced by linear independence. If the matrices $A$, $B$ are neither orthonormal nor linear independent one cannot guarantee that the state space is bounded. In particular, one may end up with only two linear independent inequalities. For instance choose $A = \text{diag}(1, -1, 1, -1)$ and $B = 2A$. This leads to the inequalities $-1 - 2y \leq x \leq 1 - 2y$ and thus defines an unbounded area in $\mathbb{R}^2$.

**Definition 5.2** For $A_1, ..., A_m \in H_n(\mathbb{C})$ traceless and orthonormal and $f : \mathbb{R}^m \to$ $H_n(\mathbb{C})$ linear, we call the set

$$\left\{ (x_1, ..., x_m) \in \mathbb{R}^m : f(x_1, ..., x_m) = \sum_{i=1}^{m} x_i A_i + \mathbb{I} \geq 0 \right\} \subset \mathbb{R}^m \qquad (5.8)$$

a $m$-section of a $n$-quantum system.

Clearly, a $m$-section of an $n$-quantum system defines a convex set in $m$ dimensions, i.e., it leads to a subset of $\mathbb{R}^m$. In particular, this map defines an $m$-dimensional manifold with corners. Following definition 5.2, one can generalize the content of Lemma 5.1 to $m$-sections. If $A_1, ..., A_m$ is a family of hermitian matrices one can find a unitary matrix $U \in \mathcal{U}(n, \mathbb{C})$ such that $U^* A_i U$ is diagonal for all $i = 1, ..., m$ if and only if they pairwise commute, i.e., $[A_i, A_j] = 0$ for all $i, j$. One can also interpret Lemma 5.1 in geometrical terms. For this notice that if $A$ and $B$ commute, one obtains $n$ affine inequalities for the possible values of $x$ and $y$. Hence they define $n$ closed half-planes, i.e., each inequality divides the $\mathbb{R}^m$ into two parts. The $m$-section is then given by the intersection of all this half-planes and therefore one obtains a closed convex polytope. Also for the case if the generators $(A_i)_{i=1}^{n}$ are anticommuting, i.e., $\{A_i, A_j\} = 2\delta_{ij}\mathbb{I}$, one can give a complete characterization of the shape of the section. In particular, the $(A_i)_{i=1}^{n}$ are a subset of the generators of a Clifford algebra. By virtue of Example 4.3, this is a section of a $n + 1$-dimensional cone. Therefore the section is always given by the $n$-dimensional ball. As a special case one reproduces that for qubits, any $\ell$-section of the state space, where $\ell \in \{1, 2, 3\}$, gives again a $\ell$-dimensional ball.

**Example 5.3** Consider a section where the generators commute. By the discussion above and Lemma 5.1, we expect that the section is given by a polytope. Let $A = \epsilon \cdot \text{diag}(1, 2, 3, -6)$ and $B = \text{diag}(2, -2, 2, -2)$ and consider $f(x, y) = xA + yB + \mathbb{I}$. As one can see in Figure 5.1, also the number of vertices is bounded by four, i.e., by the number of levels of the system.

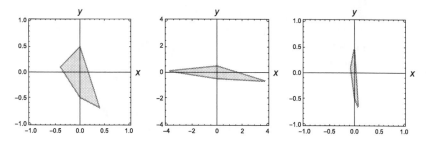

**Figure 5.1** Allowed values for $x$ and $y$ in dependence of the parameter $\epsilon \in \{0.1, 1, 5\}$

**Lemma 5.4** Let $G$ be a block diagonal matrix with blocks $A$ and $B$. Then $G \geq 0$ if and only if $A \geq 0$ and $B \geq 0$

*Proof.* Assume that $G \geq 0$. The spectrum of $G$ is given by the spectrum of $A$ and $B$. Thus $A \geq 0$ and $B \geq 0$. The equivalence of both statements follows with a similar argument. □

Unfortunately it turns out, that a polytope state space does not necessarily stem from a system of commuting matrices. In particular, to give a necessary and sufficient criterium for the shape of the state space we have to deal with two special cases. Suppose $A, B \in M_n(\mathbb{C})$ are given. First compute the set of eigenvectors of $A$ and $B$ which we will denote with $\Lambda(A)$ and $\Lambda(B)$ respectively and consider the set $\Lambda(A, B) := \Lambda(A) \cap \Lambda(B)$, which contains the mutual eigenvectors. Clearly if $A$ and $B$ not commute, the set $\Lambda(A, B)$ will not span the whole space. This leads to the following block matrices

$$A = \begin{pmatrix} A_1 & 0 \\ 0 & D_A \end{pmatrix} \quad , \quad B = \begin{pmatrix} B_1 & 0 \\ 0 & D_B \end{pmatrix} \tag{5.9}$$

where $D_A$ and $D_B$ have diagonal form. Hence the section of $A$ and $B$ reduces to

$$f(x, y) = xA + yB = \begin{pmatrix} xA_1 + yB_1 & 0 \\ 0 & xD_A + yD_B \end{pmatrix} \tag{5.10}$$

The state space defined by $f(x, y)$ can now be seen as the intersection of the convex set $\mathcal{C}$ defined by $xA_1 + yB_1$ and the polytope $\mathcal{P}$ defined by the $xD_A + yD_B$. At this point one has in principle to deal with the following problems:

1) The polytope $\mathcal{P}$ is proper contained in $\mathcal{C}$, i.e., $\mathcal{P} \subsetneq \mathcal{C}$. Then the intersection $\mathcal{P} \cap \mathcal{C}$ is $\mathcal{P}$ and thus again a polytope, while the matrices do not commute. A slightly reformulation shows that if the tupel $(x_1, ..., x_m)$ is a solution for non-commutative part, then it also has to solve the diagonal one. Thus, the non-commutative part gives no information about the shape of the state space, it is redundant.

2) One has $\mathcal{P} \not\subset \mathcal{C}$ and $\partial\mathcal{C} \cap \mathcal{P}^\circ \neq \emptyset$. Now it can happen that $\mathcal{C}$ is not given by a polytope but its boundary is flat in a certain area, i.e., there exists a parametrisation of the form

$$\Gamma := \partial\mathcal{C} \cap \mathcal{P}^\circ = \Big\{ \sum_{i=1}^{m-1} \alpha_i \mathbf{a}_i + \mathbf{b} \; : \; \alpha_j \in \mathcal{I}_j \, , \, \mathbf{a}_i, \mathbf{b} \in \mathbb{R}^m \Big\} \qquad (5.11)$$

with $\mathcal{I}_j \subset \mathbb{R}$ an interval for all $j = 1, ..., m-1$. If the sets $\mathcal{C}$ and $\mathcal{P}$ originate from a $m$-section, the set $\partial\mathcal{C} \cap \mathcal{P}^\circ$ is generically given by a $m-1$-dimensional manifold.

**Example 5.5**  Consider a 2-section of an 5-level quantum system generated by

$$f(x, y) = x \cdot \text{diag}(\epsilon, -\epsilon, 1, 2, -3) + y \begin{pmatrix} 0 & \epsilon \\ \epsilon & 0 \end{pmatrix} \oplus \text{diag}(-3, 2, 1) + \mathbb{I} \qquad (5.12)$$

Clearly $A$ and $B$ do not commute for any $\epsilon > 0$. Thus, we cannot find a basis consisting of mutual eigenvectors. In this case, one can choose the upper $2 \times 2$-block as a generator for the convex set $\mathcal{C}$. As one can see in Figure 5.2 the geometry of the section depends upon the parameter $\epsilon$. A it turns out, the section is a polytope if $\epsilon \leq \sqrt{2}$ and otherwise, it contains curvatures.

Our aim is now to classify these both cases in terms of matrices generating the section. We have already seen that $\mathcal{P}$ as well as $\mathcal{C}$ are $m$-dimensional manifolds both containing $\mathbf{0} \in \mathbb{R}^m$. Since we restrict to finite dimensions all self-adjoint operators are bounded, in particular we have that $\sigma(A_i) \subset \mathbb{R}$ for all $i = 1, ..., m$. Thus we can find $\epsilon > 0$ such that for $x_i \in (-\epsilon, \epsilon)$ we have $f(x_1, ..., x_m) \geq 0$. Hence $\mathbf{0}$ is neither at the boundary of $\mathcal{P}$ nor at the boundary of $\mathcal{C}$. This implies that there exists an open ball $\mathcal{B}_\epsilon(\mathbf{0}) \subset \mathbb{R}^m$ such that $\mathcal{B}_\epsilon(\mathbf{0}) \subset \mathcal{P} \cap \mathcal{C}$. Therefore $\mathcal{P}$ and $\mathcal{C}$ cannot intersect only at their boundaries, i.e., $\mathcal{P} \cap \mathcal{C}$ is not of measure zero with respect to the $m$-dimensional Lebesgue-measure. In the first instance we restrict to 2-sections of an $n$-quantum system.

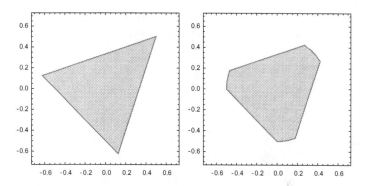

**Figure 5.2** Dependence of the geometry on the parameter $\epsilon$. On the left hand side we choose $\epsilon = 1$. On the right hand side $\epsilon$ is chosen to be 2. The section passes from a polytope to an abitrary convex set

**Theorem 5.6** ([65]). Let $A, B \in \mathbb{C}^{n \times n}$, assume that $A$ is positive semidefinite and assume that $B$ is hermitian. Then, the following statements are equivalent:

1) There exists a $\alpha \in \mathbb{C}$ such that $A + \alpha B$ is nonsingular
2) $\ker(A) \cap \ker(B) = \{0\}$
3) There exists nonzero $\alpha \in \mathbb{C}$ such that $\ker(A) \cap \ker(B + \alpha A) = \{0\}$
4) For all nonzero $\alpha \in \mathbb{C}$ one has $\ker(A) \cap \ker(B + \alpha A) = \{0\}$

**Lemma 5.7** Let $A, B, M, N \in M_n(\mathbb{C})$ with $A$ and $B$ share a common eigenvector $|\psi\rangle$. Further $A, B$ offer a decomposition of the form

$$A = a_1 M + a_2 N \quad B = b_1 M + b_2 N \tag{5.13}$$

with $(a_2 - (a_1/b_1)b_2)(a_1 - (a_2/b_2)b_1) \neq 0$. Then, also $M$ and $N$ share the common eigenvector $|\psi\rangle$.

*Proof.* We have $(a_1 M + a_2 N)|\psi\rangle = \lambda^{(1)}|\psi\rangle$ and $(b_1 M + b_2 N)|\psi\rangle = \lambda^{(2)}|\psi\rangle$. If $b_1 \neq 0$ define $\mu := -a_1/b_1$ and for $b_2 \neq 0$ put $\nu = -a_2/b_2$. Multiplying the eigenvector equation of $B$ with $\mu$ and adding to the equation of $A$ yields

$$N|\psi\rangle = \frac{\left(\mu \lambda^{(2)} + \lambda^{(1)}\right)}{a_2 + \mu b_2}|\psi\rangle =: N|\psi\rangle = \lambda^{(3)}|\psi\rangle \tag{5.14}$$

In the case of $a_2 = b_2 =$, we obtain $A = a_1 M$ and $B = b_1 N$. Thus $|\psi\rangle$ is obviously a common eigenvector of $M$ and $N$. A similar procedure leads to

$$M|\psi\rangle = \frac{\lambda^{(2)}v + \lambda^{(1)}}{a_1 + vb_1} =: \lambda^{(4)}|\psi\rangle \tag{5.15}$$

$\square$

**Lemma 5.8** Let $f = xA + yB + \mathbb{I}$ be a 2-section of $n$-level quantum system and suppose that $f$ is in common eigenvector form. Then there exists no proper interval $\mathcal{I} \subset \mathbb{R}^2$ such that $\partial \mathcal{C} \cap \mathcal{P}^\circ$ can be parametrized locally as a linear function on $\mathcal{I}$.

*Proof.* Suppose that $f$ is in the desired form, i.e.,

$$f(x, y) = \begin{pmatrix} xA_1 + yB_1 & 0 \\ 0 & xD_A + yD_B \end{pmatrix} \tag{5.16}$$

By definition of the common eigenvector form, the lower block is diagonal and therefore $[D_A, D_B] = 0$. Hence the lower block matrix defines the polytope $\mathcal{P}$. Further $D_A, D_B$ contains the maximal set of common eigenvectors of $A$ and $B$. The boundary of the convex set $\mathcal{P} \cap \mathcal{C}$ is given by the set of points where at least one of the eigenvalues of the block matrices vanishes. In terms of the characteristic polynomial this means

$$\det \begin{pmatrix} xA_1 + yB_1 & 0 \\ 0 & xD_A + yD_B \end{pmatrix} = \det(xA_1 + yB_1) \cdot \det(xD_A + yD_B) = 0 \tag{5.17}$$

Since $\Gamma = \mathcal{P}^\circ \cap \partial \mathcal{C}$ the second determinant cannot vanish for $\mathbf{x} \in \Gamma$, thus the first one must vanish. Now suppose that $\Gamma$ offers locally a linear parametrization, i.e.,

$$\Gamma = \{\alpha\mathbf{a} + \mathbf{b} \,|\, \alpha \in (-\epsilon, \epsilon)\,,\, \mathbf{a}, \mathbf{b} \in \mathbb{R}^2\} \tag{5.18}$$

Thus we have

$$\det(xA_1 + yB_1) = \det(\alpha[a_1A_1 + a_2B_1] + b_1A_1 + b_2B_1) = 0 \tag{5.19}$$

Define the new operators $E = a_1 A_1 + a_2 B_1$ and $F = b_1 A_1 + b_2 B_1$. Then (5.19) take the form $\det(\alpha E + F) = 0$ for all $\alpha \in (-\epsilon, \epsilon)$. For $\mathbf{x} \in \Gamma$ it is especially true that $x \in \partial C$. Hence $(\alpha a_1 + b_1) A_1 + (\alpha a_2 + b_2) B_1 \geq 0$ and therefore $\alpha E + F \geq 0$. Since $\alpha \in (-\epsilon, \epsilon)$ we can chose $\epsilon = 0$ and obtain $F \geq 0$. Since $\epsilon$ is taken from a continuous set, a rescaling argument leads to $E \geq 0$. Then Theorem 5.6 tells us that $E$, $F$ have a common eigenvector. By virtue of Lemma 5.7 also $A_1$, $B_1$ have a common eigenvector. But by assumption all common eigenvectors of $A_1$ and $B_1$ are stored in the matrices $D_A$ and $D_B$. Contradiction.                                    $\square$

**Theorem 5.9** Consider a 2-section of a $n$-level quantum system induced by $f$. Further assume that $\mathcal{P} \not\subset C$ and $\Gamma$ nonempty. Then $\mathcal{P} \cap C$ is not a polytope.

*Proof.* Suppose that $\partial C \cap \mathcal{P}^\circ \neq \emptyset$, $\mathcal{P} \not\subset C$ and that $\mathcal{P} \cap C$ is given by a polytope. First we show that under these assumptions $\Gamma$ always contribute to the boundary of the resulting polytope, i.e., $\partial C \cap \mathcal{P}^\circ \subset \partial (C \cap \mathcal{P})$. Since $\mathcal{P}, C \subset \mathbb{R}^2$ this follows from

$$\partial C \cap \mathcal{P}^\circ = \partial C \cap (\mathcal{P} \setminus \partial \mathcal{P}) = \partial C \cap \mathcal{P} \setminus \partial C \cap \partial \mathcal{P} \subset (\partial C \cap \mathcal{P}) \cup (C \cap \partial \mathcal{P}) = \partial (C \cap \mathcal{P})$$
$$(5.20)$$

Since $\mathcal{P} \cap C$ is a polytope, $\partial C \cap \mathcal{P}^\circ$ must be given by a finite set of points or must offer a linear parametrization. Assume first that $\partial C \cap \mathcal{P}^\circ$ is given by a finite set of points. But $C$, $\mathcal{P}$ are convex sets and $\partial C$, $\mathcal{P}^\circ$ are topologically connected. Thus, also $\partial C \cap \mathcal{P}^\circ$ are topologically connected what contradicts that there are only finitely many points. Now assume that $\partial C \cap \mathcal{P}^\circ$ offers a linear parametrization, i.e.,

$$\Gamma = \alpha \cdot \mathbf{a} + \mathbf{b} \quad \text{where} \quad \alpha \in (-\epsilon, \epsilon), \, \epsilon > 0 \qquad (5.21)$$

But this case is excluded by virtue of Lemma 5.8. Combining the two cases, one sees that this is a contradiction to the assumption that $\mathcal{P} \cap C$ is a polytope while $\Gamma$ is nonempty.                                    $\square$

The content of Lemma 5.8 and Theorem 5.9 can also be seen as a recipe of computing a section. This can be done since we can treat the convex set $C$ and the polytope $\mathcal{P}$ separately. The geometrical meaning is depicted in Figure 5.3

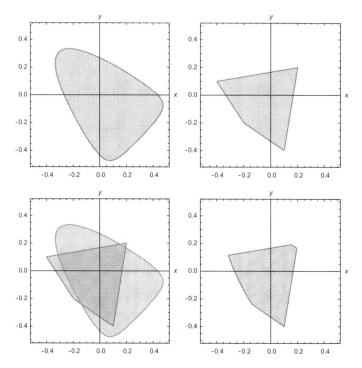

**Figure 5.3** Two dimensional section of a seven level quantum system. The first picture describes the block matrix without a common eigenvector while the second illustrates the commuting part. The whole section is given by the section of both

## 5.2    Geometric Evolution of State Spaces

So far we have just discussed the shape of sections without evolution of the system. In general the evolution is given in terms of processes, i.e., by completely positive and trace preserving maps. In particular the time evolution induced by a Hamiltonian is just a special case. By Stones-Theorem on strongly continuous unitary one-parameter groups [66], any Hamiltonian leads to a unitary group via $\mathcal{U}_t = \exp(-itH/\hbar)$ and any unitary group can be written as an exponential of a self-adjoint operator. A unitary channel $\mathcal{E}$ can thus be written as

$$\mathcal{E}_t(\rho) = U_t^* \rho U_t \tag{5.22}$$

Since $\sigma(\rho) = \sigma\left(U_t^* \rho U_t\right)$ it follows that sections are invariant under the action of unitary processes. But also the converse is true, that is if the section is invariant under a process then the process is unitary, i.e., induced by a Hamiltonian. Suppose one has a $n$-level quantum system under consideration and this system evolves in time. Since the action of the evolution operator $U_t$ affects the whole system, the question arises how the section behaves under this operation. In particular, whether sections are closed under unitary time evolution.

**Theorem 5.10** Let $H \in H_p(\mathbb{C})$ be a Hamiltonian of rank $\ell$ with pairwise distinct eigenvalues acting on a $p$-level quantum system and denote by $U_t$ the unitary one-parameter group generated by $H$. Then for $\rho \in \Omega_\ell$

$$\dim \mathrm{Orb}(\rho) = \dim\left\{\omega \in \Omega_n \mid \exists U \in U_t : \omega = U^* \rho U\right\} = \ell(\ell + 1) \tag{5.23}$$

where the dimension is the dimension of $\mathrm{Orb}(\rho)$ as an affine space.

*Proof.* Let $H$ be a Hamiltonian of rank $\ell$, i.e.,

$$H = \sum_{k=1}^{\ell} m_k |m_k\rangle\langle m_k| \tag{5.24}$$

This Hamiltonian induces a time evolution on a physical system, described by the density matrix $\rho_0$. The evolution is given by

$$\rho_t = \mathcal{E}_t(\rho_0) = U_t^* \rho_0 U_t = e^{-\frac{i}{\hbar}Ht} \rho_0 e^{\frac{i}{\hbar}Ht} \tag{5.25}$$

Further we assume that $\langle m_k | m_i \rangle = \delta_{ki}$. Thus we can write for one of the exponentials

$$e^{\frac{i}{\hbar}Ht} = \exp\left(\frac{i}{\hbar}t\left(\sum_{k=1}^{\ell} m_k |m_k\rangle\langle m_k|\right)\right) = \sum_{n=0}^{\infty}\left(\frac{i}{\hbar}\right)^n \left[\sum_{k=1}^{\ell} m_k |m_k\rangle\langle m_k|\right]^k \tag{5.26}$$

$$= \mathbb{I} + \sum_{n=1}^{\infty}\left(\frac{i}{\hbar}t\right)^n \left[\sum_{k=1}^{\ell} m_k |m_k\rangle\langle m_k|\right]^n = \mathbb{I} + \sum_{n=1}^{\infty}\sum_{k=1}^{\ell}\left(\frac{i}{\hbar}t m_k\right)^n |m_k\rangle\langle m_k|$$

$$= \mathbb{I} + \sum_{k=1}^{\ell} \left( \sum_{n=0}^{\infty} \left( \frac{i}{\hbar} t m_k \right)^n - 1 \right) |m_k\rangle\langle m_k| = \mathbb{I} - \sum_{k=1}^{\ell} |m_k\rangle\langle m_k| + \sum_{k=1}^{\ell} e^{\frac{i}{\hbar} t m_k} |m_k\rangle\langle m_k|$$

With the definitions

$$\Pi_{m_1}^{m_\ell} := \mathbb{I} - \sum_{k=1}^{\ell} |m_k\rangle\langle m_k| \quad \text{and} \quad \mathcal{P}_t\left(m_j\right) := e^{\frac{i}{\hbar} t} m_k \tag{5.27}$$

we obtain

$$e^{\frac{i}{\hbar} H t} = \Pi_{m_1}^{m_\ell} + \sum_{k=1}^{\ell} \mathcal{P}_t\left(m_k\right) |m_k\rangle\langle m_k| \tag{5.28}$$

Hence we obtain for the time evolution for the state $\rho_0$

$$\rho_t = \left( \Pi_{m_1}^{m_\ell} \rho_0 + \sum_{i=1}^{\ell} \mathcal{P}_t\left(m_i\right) |m_i\rangle\langle m_i|\rho_0 \right) \left( \Pi_{m_1}^{m_\ell} + \sum_{k=1}^{\ell} \mathcal{P}_t\left(-m_k\right) |m_k\rangle\langle m_k| \right) \tag{5.29}$$

This gives $(\ell + 1)^2$ terms and $(\ell + 1)$ of them are time independent. Thus, one ends up with $(\ell + 1)^2 - (\ell + 1) = \ell(\ell + 1)$ time dependent terms. With

$$\Pi_{m_1}^{m_\ell} = \mathbb{I} - \sum_{k=1}^{\ell} |m_k\rangle\langle m_k| = \sum_{k=\ell+1}^{p} |m_k\rangle\langle m_k| \tag{5.30}$$

and the definitions $E_{nk} := |m_n\rangle\langle m_k|$ and $\langle m_n|\rho|m_k\rangle =: \rho_{nk}$ we can write

$$\sum_{k=\ell+1}^{p} \sum_{n=1}^{\ell} \mathcal{P}_t\left(-m_n\right) E_{kn}\rho_{kn} + \sum_{n,j=1; n\neq j}^{\ell} \mathcal{P}_t\left(m_j - m_n\right) E_{jn}\rho_{jn} \tag{5.31}$$

for the time dependent part. Since the time independent part just gives an affine shift and the $m_i$ are by assumption mutual distinct the claim follows. $\qquad \square$

**Definition 5.11** Let $f : \mathbb{R}^m \to \mathrm{H}_n(\mathbb{C})$ be a section and suppose that $\mathcal{E} : \mathrm{H}_n(\mathbb{C}) \to \mathrm{H}_n(\mathbb{C})$ is a process. We define the action of $\mathcal{E}$ on the section $f$ as

$$f \mapsto \mathcal{E}\left(f\right) = \sum_{i=1}^{m} x_i \cdot \mathcal{E}\left(A_i\right) + \mathcal{E}\left(\mathbb{I}\right) \tag{5.32}$$

**Example 5.12** Let $f : \mathbb{R}^2 \to M_3\left(\mathbb{C}\right)$ with $(x, y) \mapsto xA + yB + \mathbb{I}$ where $A = \mathrm{diag}\left(1, 2 - 3\right)$ and $B = \mathrm{diag}\left(-4, 6, -2\right)$. Clearly $A$, $B$ commuting, thus the set $\{(x, y) \in \mathbb{R}^2 \mid f\left(x, y\right) \geq 0\}$ defines by Lemma 5.1 a polytope. Now consider the process $\mathcal{E}$ given by the Kraus operators

$$A_1 = \Pi_1 = \frac{1}{3}\begin{pmatrix} 1 & 1 & 1 \\ 1 & 1 & 1 \\ 1 & 1 & 1 \end{pmatrix} \ , \ A_2 = \mathbb{I} - \Pi_1 \tag{5.33}$$

Now apply the process to the section $f$, i.e.,

$$f \mapsto \mathcal{E}\left(f\right) = x \cdot \mathcal{E}\left(A\right) + y \cdot \mathcal{E}\left(B\right) + \mathcal{E}\left(\mathbb{I}\right) \tag{5.34}$$

As it turns out the commutation relation between $A,B$ changes if they undergo a process, i.e., $[\mathcal{E}\left(A\right), \mathcal{E}\left(B\right)] \neq 0$. The change of the shape of the section is depicted in Figure 5.4. Since the process is described by a projector valued measure (PVM), this example also shows that global commuting operators not necessarily commute on subspaces.

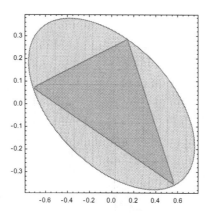

**Figure 5.4** Change of the shape of the section under a process. The triangle describes the section before the process, i.e., the generators of the section commute. The ellipse is the section after the process. Clearly the eigenvalues are not affine functions in the variables $x, y$, thus the generators cannot commute

# Conclusion

6

In this thesis we gave an introduction into the mathematical framework of order
unit spaces and their analytical and algebraic properties. This includes a discussion of the procedure of Archimedeanization and functional representations. Afterwards we used this mathematical objects to formulate the framework of generalized
probabilistic theories. This structures allow us to describe a broad range of physical theories and turned out to be a fruitful basis for fundamental investigations
in physics. We summarized the description of a single system within this theory,
applied it to classical probability theory, quantum mechanics and the PR-box and
pointed out difficulties that occur when passing over to composite systems. This
difficulties emerge in the closely connected notion of subsystems and result in different approaches to equip convex operational theories with this structure [67, 21].
The concept of sections [21] can be seen as a generalization of subsystems and
provide relations between GPTs. In particular they describe situations in which
only a subset of all observables can be implemented experimental. In Chapter 4 we
investigated the physical and algebraic properties and elaborated differences to the
algebraic framework of quantum mechanics where physical systems are described
by von-Neumann algebras. This includes an analysis of the category theoretic structure and the transformation properties of the state space. Chapter 5 formalized this
concept for the special GPT of quantum mechanics in such a way, that we obtained
a useful representation in terms of a real manifold. This allowed us to introduce
a method to characterize the geometries of two dimensional sections of an $\ell$-level
quantum system based on the theory of matrix pencils. To obtain an estimation
for the flexibility of section apart from the two dimensional case, we presented a
special hidden-variable model, called Spekkens' toy model which mimics many of
the features of a quantum bit. We phrased the Spekkens' model as a GPT, studied
the existence of such sections and gave upper and lower bounds on the dimension

J. Steinberg, *Extensions and Restrictions of Generalized Probabilistic Theories*,
BestMasters, https://doi.org/10.1007/978-3-658-37581-2_6

of the quantum system. At last we defined the action of processes on sections and showed that the geometry of the section can change under processes.

In this context, sections of qutrits are also an object of experimental interest. Those can be used to measure the size and the shape of the state space of 3-level quantum systems in analogy to [68] since they reduce the dimension of the problem. In relation with the analysis of the category theoretic structure the question arises what kind of functor $\mathcal{F}$ is. If $\mathcal{F}$ is given in terms of a forgetful functor the construction of the adjoint would be an interesting task. Another open question is whether the classification of the geometry can be generalized to three- or even higher dimensions and thus give a precise answer to the problem 4.29. Furthermore we only briefly discussed the topic of time evolution and processes and so it is an important issue to clarify with what kind of sections those are compatible.

# Bibliography

[1]  B. Hensen et al. Loophole-free Bell inequality violation using electron spins separated by 1.3 kilometres. *Nature*, 526, 2015.

[2]  G. Birkhoff and J. von Neumann. The logic of quantum mechanics. *Ann. Math.*, 37, 1936.

[3]  G. W. Mackey. Quantum mechanics and Hilbert space. *Am. Math. Mon.*, 64, 1957.

[4]  G. Ludwig. *Foundations of Quantum Mechanics*. Springer, 1983.

[5]  C. Piron. Axiomatique quantique. *Helv. Phys. Acta*, 37, 1964.

[6]  E. G. Beltrametti and G. Cassinelli. *The Logic of Quantum Mechanics*. Addison-Wesley, 1981.

[7]  G. Chriribella and R. W. Spekkens. *Quantum Theory: Informational Foundations and Foils*. Springer, 2016.

[8]  G. Chiribella, G. M. D'Ariono, and P. Perinotti. Informational derivation of Quantum Theory. *Phys. Rev. A*, 84, 2011.

[9]  L. Hardy. Reformulating and Reconstruction Quantum Theory. 2011.

[10]  P. A. Höhn and C. S. Wever. Quantum theory from questions. *Phys. Rev. A*, 95, 2017.

[11]  L. Hardy. Quantum theory from five reasonable axioms.

[12]  G. Chiribella, G. M. D'Ariano, and P. Perinotti. Probabilistic theories with purification. *Phys. Rev. A*, 81, 2010.

[13]  L. Hardy. A formalism-local framework for general probabilistic theories including quantum theory. *Math. Struct. Comput. Sci*, 23, 2013.

[14]  S. Abramsky and B. Coecke. A categorical semantics of quantum protocols. *IEEE*, 2004.

[15]  P. Seilinger. Finite dimensional Hilbert spaces are complete for dagger compact closed categories. *Log. Methods Comput. Sci.*, 8, 2012.

[16]  B. Dakic' and C. Brukner. *Quantum Theory and Beyond: Is Entanglement Special?*, page 365-392. Cambridge University Press, 2011.

[17]  L. Masanes and M. P. Müller. A derivation of quantum theory from physical requirements. *New J. Phys.*, 13, 2011.

[18]  M. A. Schlosshauer. *Decoherence and the Quantum-To-Classical Transition*. Springer, 2007.

[19]  W. H. Zurek. Decoherence, einselection, and the quantum origins of the classical. *Rev. Mod. Phys.*, 75, 2003.

[20]  M. Schlosshauer. Decoherence, the measurement problem, and interpretations of quantum mechanics. *Rev. Mod. Phys.*, 76, 2005.

© The Editor(s) (if applicable) and The Author(s), under exclusive license to
Springer Fachmedien Wiesbaden GmbH, part of Springer Nature 2022
J. Steinberg, *Extensions and Restrictions of Generalized Probabilistic Theories*,
BestMasters, https://doi.org/10.1007/978-3-658-37581-2

[21] M. Kleinmann, T.J. Osborne, V.B. Scholtz, and A.H. Werner. Typical Local Measurements in Generalized Probabilistic Theories: Emergence of Quantum Bipartite Correlations. *Phys. Rev. Lett.*, 110, 2013.

[22] O. Bratelli and D. W. Robinson. *Operator Algebras and Quantum Statistical Mechanics.* 1987.

[23] G. G. Emch. *Algebraic Methods in Statistical Mechanics and Quantum Field Theory.* Wiley, 1972.

[24] R. Haag. *Local Quantum Physics.* Springer, 1992.

[25] W. Thirring. *Quantum Mathematical Physics.* Springer, 2002.

[26] V. I. Paulsen and M. Tomforde. Vector Spaces with an Order Unit. *Indiana Univ. Math. J.*, 183, 2009.

[27] J. B. Conway. *A Course in Functional Analysis.* Springer, 2007.

[28] J. Lindenstrauss and L. Tzafriri. *Classical Banach spaces I.* Springer, 1977.

[29] E. M. Alfsen and F. W. Shultz. *State Space of Operator Algebras.* Birkhäuser Basel, 2001.

[30] R. V. Kadison. *A representation theory for commutative topological algebra.* Mem. Amer. Math. Soc., 1951.

[31] E. M. Alfsen. *Compact Convex Sets and Boundary Integrals.* Springer, 1971.

[32] M. H. Stone. A general theory of spectra i. *Proc. Nat. Acad. Sci. U.S.A*, 26:280, 1940.

[33] M. H. Stone. A general theory of spectra ii. *Proc. Nat. Acad. Sci. U.S.A*, 27:83, 1941.

[34] M. Krein and S. Krein. On an inner characteristic of the set of all continuous functions defined on a bicompact hausdorff space. *Acad. Sci. URSS.*, 27:427, 1940.

[35] S. Myers. Banach space of continuous functions. *Ann. of Math.*, 49:132, 1948.

[36] R. Arens and J. L. Kelley. Characterizations of the space of continuous functions over a compact hausdorff space. *Trans. Amer. Math. Soc.*, 62:499, 1947.

[37] J. Clarkson. A characterization of C-spaces. *Ann. of Math.*, 48:845, 1947.

[38] I. Gelfand. Normierte Ringe. *Recueil Math.*, 9:3, 1941.

[39] I. Gelfand and M. Neumark. On the imbedding of normed rings into the ring of operators in a Hilbert space. *Recueil Math.*, 12:197, 1943.

[40] K. H. Han. Tensor products of function systems revisited. *Positivity*, 20:235–255, 2016.

[41] D. Hilbert. Mathematical Problems. *Bull. Am. Math. Soc.*, 8(10):437–479, 1902.

[42] G. P. Barker. Theory of cones. *Linear Algebra Appl.*, 39, 1981.

[43] G. Aubrun, L. Lami, C. Palazuelos, and M. Plavala. Entangleability of Cones. 2019.

[44] T. Heinosaari and M. Ziman. *The Mathematical Language of Quantum Mechanics.* Cambridge University Press, 2012.

[45] B. Grünbaum. *Convex Polytopes.* Springer, 2003.

[46] K. Kraus. *States, Effects, and Operations.* Springer, 1983.

[47] G. Chiribella. Agents, Subsystems, and the Conservation of Information. *Entropy*, 20, 2018.

[48] E. B. Vinberg. Homogeneous cones. *Dokl. Acad. Nauk. SSSR*, 141, 1960.

[49] M. Koecher. Die Geodätischen von Positivitätsbereichen. *Math. Annalen*, 135, 1958.

[50] H. Barnum, J. Barrett, M. S. Leifer, and A. Wilce. Teleportation in General Probabilistic Theories. 1998.

[51] E. B. Davies and J. T. Lewis. An operational approach to quantum probability. *Comm. Math. Phys.*, 17, 1970.

[52] A. Holevo. Radon-Nikodym derivatives of quantum instruments. *J. Math. Phys.*, 39, 1998.

[53] J. Barrett. Information processing in generalized probabilistic theories. *Phys. Rev. A*, 75, 2007.

[54] H. Barnum, J. Barrett, A. Wilce, and M. Leifer. Cloning and Broadcasting in Generic Probabilistic Models. 2006.

[55] M. Horodecki, P. Horodecki, and R. Horodecki. Separability of mixed states: necessary and sufficient conditions. *Phys. Lett. A*, 223, 1996.

[56] S. Popescu and D. Rohrlich. Quantum Nonlocality as an Axiom. *Found. Phys.*, 24, 1993.

[57] S. Popescu. Nonlocality beyond quantum mechanics. *Nat. Phys.*, 10, 2014.

[58] P. Jordan and E. Wigner. Über das Paulische Äquivalenzverbot. *Zeitschrift für Physik*, 47, 1928.

[59] B. S. Tsirel'son. Quantum analogues of Bell inequalities. The case of spatially separated domains. *J. Sov. Math.*, 36, 1987.

[60] R. W. Spekkens. Evidence for the epistemic view of quantum states: A toy theory. *Phys. Rev. A*, 75, 2007.

[61] R. W. Spekkens. *Quasi-Quantization: Classical Statistical Theories with an Epistemic Restriction*. Springer Berlin Heidelberg, 2016.

[62] M. Kleinmann. Sequences of projective measurements in generalized probabilistic models. *J. Phys. A*, 47(45), 2014.

[63] R. T. Rockafellar. *Convex Analysis*. Princeton University Press, 1997.

[64] S. K. Goyal, B. N. Simon, R. Singh, and S. Simon. Geometry of the generalized Bloch sphere for qutrits. *Phys. Rev. A*, 49, 2016.

[65] D. S. Bernstein. *Matrix Mathematics*. Princeton University Press, 2009.

[66] G. Teschl. *Mathematical Methods in Quantum Mechanics*. American Mathematical Society, 2009.

[67] D. Schmid, J. H. Selby, E. Wolfe, R. Kunjwal, and R. W. Spekkens. The Characterization of Noncontextuality in the Framework of Generalized Probabilistic Theories. 2019.

[68] M. D. Mazurek, M. F. Pusey, K. J. Resch, and R. W. Spekkens. Experimentally bounding deviations from quantum theory in the landscape of generalized probabilistic theories. 2019.

Printed in the United States
by Baker & Taylor Publisher Services